The War on Drugs

Other Books in the Issues on Trial Series:

The War on Drugs

Sylvia Engdahl, Book Editor

GREENHAVEN PRESS
A part of Gale, Cengage Learning

GALE
CENGAGE Learning™

Detroit • New York • San Francisco • New Haven, Conn • Waterville, Maine • London

Christine Nasso, *Publisher*
Elizabeth Des Chenes, *Managing Editor*

© 2009 Greenhaven Press, a part of Gale, Cengage Learning

For more information, contact:
Greenhaven Press
27500 Drake Rd.
Farmington Hills, MI 48331-3535
Or you can visit our Internet site at gale.cengage.com.

For product information and technology assistance, contact us at

Gale Customer Support, 1-800-877-4253
For permission to use material from this text or product, submit all requests online at www.cengage.com/permissions

Further permissions questions can be emailed to permissionrequest@cengage.com

Articles in Greenhaven Press anthologies are often edited for length to meet page requirements. In addition, original titles of these works are changed to clearly present the main thesis and to explicitly indicate the author's opinion. Every effort is made to ensure that Greenhaven Press accurately reflects the original intent of the authors. Every effort has been made to trace the owners of copyrighted material.

Cover photograph reproduced by permission of David McNew/Getty Images.

LIBRARY OF CONGRESS CATALOGING-IN-PUBLICATION DATA

The war on drugs / Sylvia Engdahl, book editor.
    p. cm. -- (Issues on trial)
   Includes bibliographical references and index.
   ISBN-13: 978-0-7377-4346-3 (hardcover)
   1. Narcotic laws--United States--Criminal provisions. 2. Due process of law--United States. 3. Privacy, Right of--United States. 4. Drug abuse--United States--Prevention. I. Engdahl, Sylvia.
   KF3890.W37 2009
   345.73'0277--dc22

                                  2008049404

Printed in the United States of America
1 2 3 4 5 6 7 13 12 11 10 09

Contents

Chapter 1: Drug Testing of Government Employees Is Not Unconstitutional

Chapter 3: Drug Testing of Participants in Extracurricular Activities Is Legal

Chapter 4: The Federal Government Can Prohibit Medical Use of Marijuana

Foreword

The U.S. courts have long served as a battleground for the most highly charged and contentious issues of the time. Divisive matters are often brought into the legal system by activists who feel strongly for their cause and demand an official resolution. Indeed, subjects that give rise to intense emotions or involve closely held religious or moral beliefs lay at the heart of the most polemical court rulings in history. One such case was *Brown v. Board of Education* (1954), which ended racial segregation in schools. Prior to *Brown*, the courts had held that blacks could be forced to use separate facilities as long as these facilities were equal to that of whites.

For years many groups had opposed segregation based on religious, moral, and legal grounds. Educators produced heartfelt testimony that segregated schooling greatly disadvantaged black children. They noted that in comparison to whites, blacks received a substandard education in deplorable conditions. Religious leaders such as Martin Luther King Jr. preached that the harsh treatment of blacks was immoral and unjust. Many involved in civil rights law, such as Thurgood Marshall, called for equal protection of all people under the law, as their study of the Constitution had indicated that segregation was illegal and un-American. Whatever their motivation for ending the practice, and despite the threats they received from segregationists, these ardent activists remained unwavering in their cause.

Those fighting against the integration of schools were mainly white southerners who did not believe that whites and blacks should intermingle. Blacks were subordinate to whites, they maintained, and society had to resist any attempt to break down strict color lines. Some white southerners charged that segregated schooling was *not* hindering blacks' education. For example, Virginia attorney general J. Lindsay Almond asserted, "With the help and the sympathy and the love and respect of the white people of the South, the colored man has risen under that educational process to a place of eminence and respect throughout the nation. It has served him well." So when the Supreme Court ruled against the segregationists in *Brown*, the South responded with vociferous cries of protest. Even government leaders criticized the decision. The governor of Arkansas, Orval Faubus, stated that he would not "be a party to any attempt to force acceptance of change to which the people are so overwhelmingly opposed." Indeed, resistance to integration was so great that when black students arrived at the formerly all-white Central High School in Arkansas, federal troops had to be dispatched to quell a threatening mob of protesters.

Nevertheless, the *Brown* decision was enforced and the South integrated its schools. In this instance, the Court, while not settling the issue to everyone's satisfaction, functioned as an instrument of progress by forcing a major social change. Historian David Halberstam observes that the *Brown* ruling "deprived segregationist practices of their moral legitimacy. . . . It was therefore perhaps the single most important moment of the decade, the moment that separated the old order from the new and helped create the tumultuous era just arriving." Considered one of the most important victories for civil rights, *Brown* paved the way for challenges to racial segregation in many areas, including on public buses and in restaurants.

In examining *Brown*, it becomes apparent that the courts play an influential role—and face an arduous challenge—in shaping the debate over emotionally charged social issues. Judges must balance competing interests, keeping in mind the high stakes and intense emotions on both sides. As exemplified by *Brown*, judicial decisions often upset the status quo and initiate significant changes in society. Greenhaven Press's Issues on Trial series captures the controversy surrounding influential court rulings and explores the social ramifications of such decisions from varying perspectives. Each anthology highlights one social issue—such as the death penalty, students' rights, or wartime civil liberties. Each volume then focuses on key historical and contemporary court cases that helped mold the issue as we know it today. The books include a compendium of primary sources—court rulings, dissents, and immediate reactions to the rulings—as well as secondary sources from experts in the field, people involved in the cases, legal analysts, and other commentators opining on the implications and legacy of the chosen cases. An annotated table of contents, an in-depth introduction, and prefaces that overview each case all provide context as readers delve into the topic at hand. To help students fully probe the subject, each volume contains book and periodical bibliographies, a comprehensive index, and a list of organizations to contact. With these features, the Issues on Trial series offers a well-rounded perspective on the courts' role in framing society's thorniest, most impassioned debates.

Introduction

The term "war on drugs" was first used by President Richard Nixon in a 1969 message to Congress identifying drug abuse as a serious national threat. Two years later, after the Controlled Substances Act (CSA) had been passed, he officially declared a war on drugs, calling drug abuse "public enemy no. 1." After another two years, in July of 1973, he created the Drug Enforcement Administration (DEA) to coordinate the work of the agencies involved in enforcement. Today, it is taken for granted that preventing the use of harmful drugs is the responsibility of the federal government.

This was not always the case. In the nineteenth century, drugs were not controlled; narcotics were often put into patent medicines, causing patients to become unknowingly addicted. Early in the twentieth century, the Food and Drugs Act was passed to ensure the purity and honest labeling of all products sold. The first law aimed specifically at narcotics was the Harrison Act of 1914, which regulated and taxed narcotics but did not ban them. At that time, however, public opposition to the use of mind-altering substances was growing, and in 1920, the Eighteenth Amendment to the Constitution became law, prohibiting the sale, but not the private use, of alcohol. Prohibition of alcohol had unintended results. It created a black market that fostered the rise of organized crime, leading to corruption and violence that proved increasingly dangerous to society, and so in 1933 the Eighteenth Amendment was repealed. The crime bosses, deprived of their lucrative alcohol business, then turned to selling other drugs, the use of which became more widespread than it had been previously. Thus for the first time, drug abuse became a major social problem, and the stage for the war on drugs was set.

Most people agree that drugs banned by the CSA, with the possible exception of marijuana, are far more harmful than alcohol and that everything possible should be done to prevent their use. But not everyone believes that the war on drugs is an effective means of prevention. Many now feel that it is operating just as alcohol prohibition did—failing to stop consumption and, through the black market it creates, causing more violent crimes, and even more addiction, than would otherwise exist. The vast majority of the citizens who advocate legalization of drugs are as strongly opposed to their use as those who support prohibition. They argue, however, that a better way of reducing drug use would be to take the profit out of drug-related crimes.

Moreover, opponents of the war on drugs point out that many innocent people are its victims, and that citizens in general—both the innocent and those guilty of minor drug offenses—have lost some of their constitutional rights. Drug issues involving constitutional rights have been brought before the Supreme Court, and almost always, the Court has ruled in favor of increasing government power to stamp out drug use at the cost of allowing otherwise established rights to be compromised.

Even the right to free speech has been abridged. Attorney Alex Coolman wrote in his Drug Law blog, "This year [2007] the Supreme Court gave us *Morse v. Frederick* and the 'Drug Exception' to the First Amendment, another in a series of decisions that erode constitutional protections for the sake of the war on drugs. . . . The Supreme Court concluded that it was just fine for schools to create content-based restrictions on student speech if 'a reasonable observer would interpret [that speech] as advocating illegal drug use.' The Bong Hits case [which involved a student banner that contained an apparent reference to marijuana] is just one more sad illustration of the way the drug war is chipping away at the American Constitution."

The cases covered in *Issues on Trial: The War on Drugs* deal with other provisions of the Constitution. Does testing of employees or students for illegal drug use constitute a violation of the Fourth Amendment's ban on unreasonable searches? The Supreme Court said it does not. Does the Constitution's Commerce Clause give the federal government the power to override state laws that permit medical use of marijuana? Yes, according to the Supreme Court majority. Can the government seize the property of someone who is sent to prison for drug crimes, even though the Fifth Amendment does not allow double punishment for the same offense? Yes, said the Court, although in an earlier case, *United States v. Lasanta* (1992), the U. S. Court of Appeals for the Second Circuit stated, "While Congress may have tended civil forfeiture to be a 'powerful weapon in the war on drugs,' it would, indeed, be a Pyrrhic victory [a victory more costly than what it achieves] for the country, if the government's relentless and imaginative use of that weapon were to leave the Constitution itself a casualty."

This is not to say that the Supreme Court invariably supports drug war measures; in *Chandler v. Miller* (1997) it struck down a Georgia state law requiring political candidates to submit to urine tests for illegal drugs as a condition for appearing on the ballot. Furthermore, its opinions are rarely unanimous, and the dissenting justices have often expressed strong arguments against the majority decisions. Balancing two important principles against each other is never easy. No one wants drug abuse to continue, nor does anyone want criminals to go on profiting from it. Whether the war on drugs can be won without destroying too many civil liberties will remain an area of controversy.

Drug Testing of Government Employees Is Not Unconstitutional

Case Overview

National Treasury Employees Union v. Von Raab (1989)

The screening of employees for illegal drug use is now routine in many industries. But this was not always the case. The first screening of people not individually suspected of drug use took place in 1970 when, after a tragic shipboard accident due to drug-induced impairment, the U.S. Navy began testing all active duty personnel. Testing of employees by large corporations began in the early 1980s. However, many civilian programs were struck down by the courts, which generally held that suspicionless testing was unconstitutional under the Fourth Amendment's protection against unreasonable searches.

In 1986, President Ronald Reagan issued an Executive Order declaring illegal drug use unacceptable for federal employees and directing federal agencies to develop plans for testing job applicants and employees in sensitive positions. Congress then passed the Drug-Free Workplace Act, which required companies with government contracts to maintain a drug-free workplace, and courts became more sympathetic to routine drug testing.

One of the first federal agencies to establish a drug testing program was the United States Customs Service, a bureau of the Department of the Treasury responsible for processing persons, cargo, and mail into the United States and enforcing customs and related laws. An important function of Customs is the seizure of contraband, including illegal drugs. In May of 1986 the Commissioner of Customs, William von Raab, announced that urine testing would be required of applicants for positions that were directly involved in drug interdiction or enforcement of related laws; that required carrying firearms; or that included access to classified material—which he feared might fall into the hands of drug smugglers if drug-using employees were blackmailed. The National Treasury Employees Union, representing current employees who sought promotion to such positions, filed suit on the grounds that the proposed program was unconstitutional.

The District Court ruled in the employees' favor and ordered the Customs Service not to require the tests, but the Court of Appeals allowed testing to proceed. It held that although drug screening does constitute a search within the meaning of the Fourth Amendment, in this case the search was reasonable because of the government's strong interest in detecting drug use among personnel responsible for enforcing drug laws. The union then appealed to the U.S. Supreme Court. In a five to four ruling, the Supreme Court upheld the testing of employees who were directly involved in drug interdiction or who were required to carry firearms. It decided, however, that the Customs Service had included too many positions among those said to involve access to high-level classified material, and sent that aspect of the case back to the lower court to be further investigated.

On the same day as its decision in *National Treasury Employees Union v. Von Raab*, the Supreme Court decided a related case, *Skinner v. Railway Labor Executives' Association*, which dealt with mandatory testing of railroad employees involved in train accidents. Although because safety was at issue, more justices felt drug testing was justified than approved of it in *Von Raab*, two of them maintained that constitutional protections should not be overridden. In his dissenting opinion, Justice Thurgood Marshall wrote:

> The issue in this case is not whether declaring a war on illegal drugs is good public policy. The importance of ridding our society of such drugs is, by now, apparent to all. Rather, the issue here is whether the Government's deployment in that war of a particularly draconian weapon—the compul-

sory collection and chemical testing of railroad workers' blood and urine—comports with the Fourth Amendment. Precisely because the need for action against the drug scourge is manifest, the need for vigilance against unconstitutional excess is great. History teaches that grave threats to liberty often come in times of urgency, when constitutional rights seem too extravagant to endure. . . . There is no drug exception to the Constitution.

In *National Treasury Employees Union v. Von Raab*, Justice Antonin Scalia made similar comments in his dissent, which is included in this chapter. However, many people believe that it is less important for policies to preserve the rights of employees who do not use drugs than to find the small minority who do. Since the *Skinner* and *Von Raab* rulings, suspicionless drug testing has been widely accepted.

> "The Government's compelling interests in preventing the promotion of drug users to positions where they might endanger the integrity of our Nation's borders or the life of the citizenry outweigh the privacy interests of those who seek promotion to these positions."

The Court's Decision: Routine Drug Testing Does Not Violate the Fourth Amendment

Anthony M. Kennedy

Anthony M. Kennedy has been a justice of the Supreme Court since 1988. He has acted as the Court's swing vote between liberals and conservatives in many five–four decisions. In the following, he explains why the Court upheld drug screening of Customs Service employees despite the fact that they were not under suspicion and few, if any, tested positive for drugs. The Court agreed that the Fourth Amendment protects government employees from unreasonable searches, he says, and normally searches require both a warrant and probable cause. However, there are exceptions. In this case, the purpose of the testing was not to catch criminals, and in fact, the results could not be used for criminal prosecution. Rather, it aimed to deter drug use among employees eligible for promotion to sensitive positions involving detection of drug smuggling, where users might be unsympathetic to their mission and in any case would be at risk for bribery and blackmail. The public interest also demands that drug

Anthony M. Kennedy, majority opinion, *National Treasury Employees Union v. Von Raab*, U.S. Supreme Court, March 21, 1989.

21

users be barred from positions that require carrying firearms. However, it was not clear to the Court whether screening of all employees promoted to positions where they might handle classified information was reasonable, so it sent the latter issue back to the lower court for further investigation.

In *Skinner v. Railway Labor Executives' Assn.*, decided today [March 21, 1989], we held that federal regulations requiring employees of private railroads to produce urine samples for chemical testing implicate the Fourth Amendment, as those tests invade reasonable expectations of privacy. Our earlier cases have settled that the Fourth Amendment protects individuals from unreasonable searches conducted by the Government, even when the Government acts as an employer, and, in view of our holding in *Railway Labor Executives* that urine tests are searches, it follows that the Customs Service's drug testing program must meet the reasonableness requirement of the Fourth Amendment.

While we have often emphasized, and reiterate today, that a search must be supported, as a general matter, by a warrant issued upon probable cause, our decision in *Railway Labor Executives* reaffirms the longstanding principle that neither a warrant nor probable cause, nor, indeed, any measure of individualized suspicion, is an indispensable component of reasonableness in every circumstance. As we note in *Railway Labor Executives*, our cases establish that, where a Fourth Amendment intrusion serves special governmental needs beyond the normal need for law enforcement, it is necessary to balance the individual's privacy expectations against the Government's interests to determine whether it is impractical to require a warrant or some level of individualized suspicion in the particular context.

It is clear that the Customs Service's drug testing program is not designed to serve the ordinary needs of law enforcement. Test results may not be used in a criminal prosecution of the employee without the employee's consent. The purposes of the program are to deter drug use among those eligible for promotion to sensitive positions within the Service, and to prevent the promotion of drug users to those positions. These substantial interests, no less than the Government's concern for safe rail transportation at issue in *Railway Labor Executives*, present a special need that may justify departure from the ordinary warrant and probable cause requirements.

Not All Violations of Privacy Require Warrants

Petitioners [the National Treasury Employees Union] do not contend that a warrant is required by the balance of privacy and governmental interests in this context, nor could any such contention withstand scrutiny. We have recognized before that requiring the Government to procure a warrant for every work-related intrusion would conflict with "the common sense realization that government offices could not function if every employment decision became a constitutional matter" [*O'Connor v. Ortega* (1983)]. Even if Customs Service employees are more likely to be familiar with the procedures required to obtain a warrant than most other Government workers, requiring a warrant in this context would serve only to divert valuable agency resources from the Service's primary mission. The Customs Service has been entrusted with pressing responsibilities, and its mission would be compromised if it were required to seek search warrants in connection with routine, yet sensitive, employment decisions.

Furthermore, a warrant would provide little or nothing in the way of additional protection of personal privacy. A warrant serves primarily to advise the citizen that an intrusion is authorized by law and limited in its permissible scope, and to interpose a neutral magistrate between the citizen and the law enforcement officer "engaged in the often competitive enterprise of ferreting out crime" [*Johnson v. United States* (1948)].

But, in the present context the circumstances justifying toxicological testing and the permissible limits of such intrusions are defined narrowly and specifically . . . and doubtless are well-known to covered employees. Under the Customs program, every employee who seeks a transfer to a covered position knows that he must take a drug test, and is likewise aware of the procedures the Service must follow in administering the test. . . .

Customs Officers Have Contact with Drug Traffickers

Even where it is reasonable to dispense with the warrant requirement in the particular circumstances, a search ordinarily must be based on probable cause. Our cases teach, however, that the probable cause standard "'is peculiarly related to criminal investigations'" [*Colorado v. Bertine* (1987)]. In particular, the traditional probable cause standard may be unhelpful in analyzing the reasonableness of routine administrative functions, especially where the Government seeks to prevent the development of hazardous conditions or to detect violations that rarely generate articulable grounds for searching any particular place or person. Our precedents have settled that, in certain limited circumstances, the Government's need to discover such latent or hidden conditions, or to prevent their development, is sufficiently compelling to justify the intrusion on privacy entailed by conducting such searches without any measure of individualized suspicion. We think the Government's need to conduct the suspicionless searches required by the Customs program outweighs the privacy interests of employees engaged directly in drug interdiction, and of those who otherwise are required to carry firearms.

The Customs Service is our Nation's first line of defense against one of the greatest problems affecting the health and welfare of our population. We have adverted before to "the veritable national crisis in law enforcement caused by smuggling of illicit narcotics" [*United States v. Montoya de Hernandez* (1985)]. Our cases also reflect the trafficker's seemingly inexhaustible repertoire of deceptive practices and elaborate schemes for importing narcotics. . . . The record in this case confirms that, through the adroit selection of source locations, smuggling routes, and increasingly elaborate methods of concealment, drug traffickers have managed to bring into this country increasingly large quantities of illegal drugs. The record also indicates, and it is well-known, that drug smugglers do not hesitate to use violence to protect their lucrative trade and avoid apprehension.

Many of the Service's employees are often exposed to this criminal element and to the controlled substances it seeks to smuggle into the country. The physical safety of these employees may be threatened, and many may be tempted not only by bribes from the traffickers with whom they deal, but also by their own access to vast sources of valuable contraband seized and controlled by the Service. The Commissioner [Von Raab, the respondent in the case] indicated below that "Customs [o]fficers have been shot, stabbed, run over, dragged by automobiles, and assaulted with blunt objects while performing their duties." At least nine officers have died in the line of duty since 1974. He also noted that Customs officers have been the targets of bribery by drug smugglers on numerous occasions, and several have been removed from the Service for accepting bribes and for other integrity violations. . . .

It is readily apparent that the Government has a compelling interest in ensuring that frontline interdiction personnel are physically fit, and have unimpeachable integrity and judgment. Indeed, the Government's interest here is at least as important as its interest in searching travelers entering the country. We have long held that travelers seeking to enter the country may be stopped and required to submit to a routine search without probable cause, or even founded suspicion, "because of national self-protection reasonably requiring one entering the country to identify himself as entitled to come in, and his belongings as effects which may be lawfully brought in" [*Carroll v. United States* (1925)]. This national interest in self-protection could be irreparably damaged if those charged with safeguarding it were, because of their own drug use, unsympathetic to their mission of interdicting narcotics. A drug user's indifference to the Service's basic mission or, even worse, his active complicity with the malefactors, can facilitate importation of sizable drug shipments or block apprehension of dangerous criminals. The public interest demands effective measures to bar drug users from positions directly involving the interdiction of illegal drugs.

Drug Users Must Be Barred from Positions Where They Carry Firearms

The public interest likewise demands effective measures to prevent the promotion of drug users to positions that require the incumbent to carry a firearm, even if the incumbent is not engaged directly in the interdiction of drugs. Customs employees who may use deadly force plainly "discharge duties fraught with such risks of injury to others that even a momentary lapse of attention can have disastrous consequences." We agree with the Government that the public should not bear the risk that employees who may suffer from impaired perception and judgment will be promoted to positions where they may need to employ deadly force. Indeed, ensuring against the creation of this dangerous risk will itself further Fourth Amendment values, as the use of deadly force may violate the Fourth Amendment in certain circumstances.

Against these valid public interests we must weigh the interference with individual liberty that results from requiring these classes of employees to undergo a urine test. The interference with individual privacy that results from the collection of a urine sample for subsequent chemical analysis could be substantial in some circumstances. We have recognized, however, that the "operational realities of the workplace" may render entirely reasonable certain work-related intrusions by supervisors and coworkers that might be viewed as unreasonable in other contexts. While these operational realities will rarely affect an employee's expectations of privacy with respect to searches of his person, or of personal effects that the employee may bring to the workplace, it is plain that certain forms of public employment may diminish privacy expectations even with respect to such personal searches. Employees of the United States Mint, for example, should expect to be subject to certain routine personal searches when they leave the workplace every day. Similarly, those who join our military or intelligence services may not only be required to give what in other contexts might be viewed as extraordinary assurances of trustworthiness and probity, but also may expect intrusive inquiries into their physical fitness for those special positions.

We think Customs employees who are directly involved in the interdiction of illegal drugs or who are required to carry firearms in the line of duty likewise have a diminished expectation of privacy in respect to the intrusions occasioned by a urine test. Unlike most private citizens or government employees in general, employees involved in drug interdiction reasonably should expect effective inquiry into their fitness and probity. Much the same is true of employees who are required to carry firearms. Because successful performance of their duties depends uniquely on their judgment and dexterity, these employees cannot reasonably expect to keep from the Service personal information that bears directly on their fitness. While reasonable tests designed to elicit this information doubtless infringe some privacy expectations, we do not believe these expectations outweigh the Government's compelling interests in safety and in the integrity of our borders.

The Testing Program Is Reasonable Despite Revealing Little Drug Use

Without disparaging the importance of the governmental interests that support the suspicionless searches of these employees, petitioners nevertheless contend that the Service's drug testing program is unreasonable in two particulars. First, petitioners argue that the program is unjustified because it is not based on a belief that testing will reveal any drug use by covered employees. In pressing this argument, petitioners point out that the Service's testing scheme was not implemented in response to any perceived drug problem among Customs employees, and that the program actually has not led to the discovery of a significant number of drug users. Counsel for petitioners informed us at oral argument that no more than 5 employees out of 3,600 have tested positive for drugs. Second, petitioners contend that the Service's scheme is not a "sufficiently productive mechanism to justify [its] intrusion upon Fourth Amendment interests" [*Delaware v. Prouse* (1979)] because illegal drug users can avoid detection with ease by temporary abstinence or by surreptitious adulteration of their urine specimens. These contentions are unpersuasive.

Petitioners' first contention evinces an unduly narrow view of the context in which the Service's testing program was implemented. Petitioners do not dispute, nor can there be doubt, that drug abuse is one of the most serious problems confronting our society today. There is little reason to believe that American workplaces are immune from this pervasive social problem, as is amply illustrated by our decision in *Railway Labor Executives*. Detecting drug impairment on the part of employees can be a difficult task, especially where, as here, it is not feasible to subject employees and their work product to the kind of day-to-day scrutiny that is the norm in more traditional office environments. Indeed, the almost unique mission of the Service gives the Government a compelling interest in ensuring that many of these covered employees do not use drugs even off duty, for such use creates risks of bribery and blackmail against which the Government is entitled to guard. In light of the extraordinary safety and national security hazards that would attend the promotion of drug users to positions that require the carrying of firearms or the interdiction of controlled substances, the Service's policy of deterring drug users from seeking such promotions cannot be deemed unreasonable.

The mere circumstance that all but a few of the employees tested are entirely innocent of wrongdoing does not impugn the program's validity. The same is likely to be true of householders who are required to submit to suspicionless housing code inspections, and of motorists who are stopped at the checkpoints we approved in *United States v. Martinez-Fuerte* (1976). The Service's program is designed to prevent the promotion of drug users to sensitive positions as much as it is designed to detect those employees who use drugs. Where, as here, the possible harm against which the Government seeks to guard is substantial, the need to prevent its occurrence furnishes an ample justification for reasonable searches calculated to advance the Government's goal.

We think petitioners' second argument—that the Service's testing program is ineffective because employees may attempt to deceive the test by a brief abstention before the test date, or by adulterating their urine specimens—overstates the case. As the Court of Appeals noted, addicts may be unable to abstain even for a limited period of time, or may be unaware of the "fade-away effect" of certain drugs. More importantly, the avoidance techniques suggested by petitioners are fraught with uncertainty and risks for those employees who venture to attempt them. A particular employee's pattern of elimination for a given drug cannot be predicted with perfect accuracy, and, in any event, this information is not likely to be known or available to the employee. Petitioners' own expert indicated below that the time it takes for particular drugs to become undetectable in urine can vary widely depending on the individual, and may extend for as long as 22 days. Thus, contrary to petitioners' suggestion, no employee reasonably can expect to deceive the test by the simple expedient of abstaining after the test date is assigned. Nor can he expect attempts at adulteration to succeed, in view of the precautions taken by the sample collector to ensure the integrity of the sample. In all the circumstances, we are persuaded that the program bears a close and substantial relation to the Service's goal of deterring drug users from seeking promotion to sensitive positions.

In sum, we believe the Government has demonstrated that its compelling interests in safeguarding our borders and the public safety outweigh the privacy expectations of employees who seek to be promoted to positions that directly involve the interdiction of illegal drugs or that require the incumbent to carry a firearm. We hold that the testing of these employees is reasonable under the Fourth Amendment.

We are unable, on the present record, to assess the reasonableness of the Government's testing program insofar as it covers employees who are required to "handle classified material." We readily agree that the Government has a compelling interest in protecting truly sensitive information from those who, "under compulsion of circumstances or for other reasons . . . might compromise [such] information" [*Department of Navy v. Egan* (1968)]. We also agree that employees who seek promotions to positions where they would handle sensitive information can be required to submit to a urine test under the Service's screening program, especially if the positions covered under this category require background investigations, medical examinations, or other intrusions that may be expected to diminish their expectations of privacy in respect of a urinalysis test.

It is not clear, however, whether the category defined by the Service's testing directive encompasses only those Customs employees likely to gain access to sensitive information. . . .

We hold that the suspicionless testing of employees who apply for promotion to positions directly involving the interdiction of illegal drugs, or to positions that require the incumbent to carry a firearm, is reasonable. The Government's compelling interests in preventing the promotion of drug users to positions where they might endanger the integrity of our Nation's borders or the life of the citizenry outweigh the privacy interests of those who seek promotion to these positions, who enjoy a diminished expectation of privacy by virtue of the special, and obvious, physical and ethical demands of those positions. We do not decide whether testing those who apply for promotion to positions where they would handle "classified" information is reasonable, because we find the record inadequate for this purpose.

> *"Symbolism, even symbolism for so worthy a cause as the abolition of unlawful drugs, cannot validate an otherwise unreasonable search."*

Dissenting Opinion: Drug Testing Without Reason for Suspicion Is Not Justifiable

Antonin Scalia

Antonin Scalia became a justice of the U.S Supreme Court in 1986, and as of 2008, he is its second most senior member. He is a strong conservative who believes in a strict interpretation of the Constitution, according to the meaning it had when originally adopted. In the following dissent in the case of National Treasury Employees Union v. Von Raab *(1989), which is often quoted by opponents of drug testing, he criticizes the Court's ruling. In the past, he says, the Court had allowed bodily searches of people not individually suspected of wrongdoing only in the case of prison inmates. He points out that whether the Fourth Amendment's prohibition of unreasonable searches applies in a case depends on what is considered "reasonable," and that this is a question of what social necessity has prompted the search. In prior cases, real problems existed whenever a widespread search of any kind was authorized. But, Scalia says, the Court's majority opinion did not present any evidence of a real problem that would be solved by drug testing of Customs Service employees, as none of the speculative ones mentioned had actually occurred. He does not believe that they are the true reason for the testing*

Antonin Scalia, dissenting opinion, *National Treasury Employees Union v. Von Raab*, U.S. Supreme Court, March 21, 1989.

program, which in his opinion, was prompted by the Government's desire to set an example and show that it is serious about the war on drugs. This, he argues, is a merely symbolic benefit, and it does not justify overriding the Fourth Amendment.

The issue in this case, [*National Treasury Employees Union v. Von Raab* (1989)] is not whether Customs Service employees can constitutionally be denied promotion, or even dismissed, for a single instance of unlawful drug use, at home or at work. They assuredly can. The issue here is what steps can constitutionally be taken to *detect* such drug use. The Government asserts it can demand that employees perform "an excretory function traditionally shielded by great privacy," while "a monitor of the same sex . . . remains close at hand to listen for the normal sounds," and that the excretion thus produced be turned over to the Government for chemical analysis. The Court agrees that this constitutes a search for purposes of the Fourth Amendment—and I think it obvious that it is a type of search particularly destructive of privacy and offensive to personal dignity.

Until today, this Court had upheld a bodily search separate from arrest and without individualized suspicion of wrongdoing only with respect to prison inmates, relying upon the uniquely dangerous nature of that environment. Today, in *Skinner* [*v. Railway Labor Executives* (1989)], we allow a less intrusive bodily search of railroad employees involved in train accidents. I joined the Court's opinion there because the demonstrated frequency of drug and alcohol use by the targeted class of employees, and the demonstrated connection between such used and grave harm, rendered the search a reasonable means of protecting society. I decline to join the Court's opinion in the present case because neither frequency of use nor connection to harm is demonstrated, or even likely. In my view, the Customs Service rules are a kind of immolation of privacy and human dignity in symbolic opposition to drug use.

The Fourth Amendment protects the "right of the people to be secure in their persons, houses, papers, and effects, against unreasonable searches and seizures." While there are some absolutes in Fourth Amendment law, as soon as those have been left behind and the question comes down to whether a particular search has been "reasonable," the answer depends largely upon the social necessity that prompts the search. Thus, in upholding the administrative search of a student's purse in a school, we began with the observation (documented by an agency report to Congress) that "[m]aintaining order in the classroom has never been easy, but in recent years, school disorder has often taken particularly ugly forms: Drug use and violent crime in the schools have become major social problems."

When we approved fixed checkpoints near the Mexican border to stop and search cars for illegal aliens, we observed at the outset that "the Immigration and Naturalization Service now suggests there may be as many as 10 or 12 million aliens illegally in the country," and that "[i]nterdicting the flow of illegal entrants from Mexico poses formidable law enforcement problems" [*United States v. Martinez-Fuerte* (1976)]. And the substantive analysis of our opinion today in *Skinner* begins, "[t]he problem of alcohol use on American railroads is as old as the industry itself," and goes on to cite statistics concerning that problem and the accidents it causes, including a 1979 study finding that "23% of the operating personnel were 'problem drinkers.'"

There Is No Evidence That Drug Testing of Customs Service Employees Will Solve a Real Problem

The Court's opinion in the present case, however, will be searched in vain for real evidence of a real problem that will be solved by urine testing of Customs Service employees. Instead, there are assurances that [t]he Customs Service is our Nation's first line of defense against one of the greatest problems affecting the health and welfare of our population, that "[m]any of the Service's employees are often exposed to [drug smugglers] and to the controlled substances they seek to smuggle into the country"; that "customs officers have been the targets of bribery by drug smugglers on numerous occasions, and several have been removed from the Service for accepting bribes and other integrity violation"; that "the Government has a compelling interest in ensuring that front-line interdiction personnel are physically fit, and have unimpeachable integrity and judgment"; that the "national interest in self-protection could be irreparably damaged if those charged with safeguarding it were, because of their own drug use, unsympathetic to their mission of interdicting narcotics"; and that "the public should not bear the risk that employees who may suffer from impaired perception and judgment will be promoted to positions where they may need to employ deadly force."

To paraphrase [Winston] Churchill, all this contains much that is obviously true, and much that is relevant; unfortunately, what is obviously true is not relevant, and what is relevant is not obviously true. The only pertinent points, it seems to me, are supported by nothing but speculation, and not very plausible speculation at that. It is not apparent to me that a Customs Service employee who uses drugs is significantly more likely to be bribed by a drug smuggler, any more than a Customs Service employee who wears diamonds is significantly more likely to be bribed by a diamond smuggler— unless, perhaps, the addiction to drugs is so severe, and requires so much money to maintain, that it would be detectable even without benefit of a urine test. Nor is it apparent to me that Customs officers who use drugs will be appreciably less "sympathetic" to their drug interdiction mission, any more than police officers who exceed the speed limit in their private cars are appreciably less sympathetic to their mission of enforcing the traffic laws. (The only difference is that the Customs officer's individual efforts, if they are irreplaceable, can theoretically affect the availability of his own drug supply—a prospect so remote as to be an absurd basis of motivation.) Nor, finally, is it apparent to me that urine tests will be even marginally more effective in preventing gun-carrying agents from risking "impaired perception and judgment" than is their current knowledge that, if impaired, they may be shot dead in unequal combat with unimpaired smugglers—unless, again, their addiction is so severe that no urine test is needed for detection.

What is absent in the Government's justifications—notably absent, revealingly absent, and, as far as I am concerned, dispositively absent—is the recitation of *even a single instance* in which any of the speculated horribles actually occurred: An instance, that is, in which the cause of bribe-taking, or of poor aim, or of unsympathetic law enforcement, or of compromise of classified information, was drug use. Although the Court points out that several employees have in the past been removed from the Service for accepting bribes and other integrity violations, and that at least nine officers have died in the line of duty since 1974, there is no indication whatever that these incidents were related to drug use by Service employees. Perhaps concrete evidence of the severity of a problem is unnecessary when it is so well-known that courts can almost take judicial notice of it; but that is surely not the case here. The Commissioner of Customs himself has stated that he "believe[s] that Customs is largely drug-free," that "[t]he extent of illegal drug use by Customs employees was not the reason for establishing this program," and that he "hope[s] and expect[s] to receive reports of very few positive findings through drug screening." The test results have fulfilled those hopes and expectations. According to the Service's counsel, out of 3,600 employees tested, no more than 5 tested positive for drugs.

The Court's response to this lack of evidence is that "[t]here is little reason to believe that American workplaces are immune from [the] pervasive social problem" of drug abuse. Perhaps such a generalization would suffice if the workplace at issue could produce such catastrophic social harm that no risk whatever is tolerable—the secured areas of a nuclear power plant, for example. But if such a generalization suffices to justify demeaning bodily searches, without particularized suspicion, to guard against the bribing or blackmailing of a law enforcement agent, or the careless use of a firearm, then the Fourth Amendment has become frail protection indeed. . . .

Today's decision would be wrong, but at least of more limited effect, if its approval of drug testing were confined to that category of employees assigned specifically to drug interdiction duties. Relatively few public employees fit that description. But in extending approval of drug testing to that category consisting of employees who carry firearms, the Court exposes vast numbers of public employees to this needless indignity. Logically, of course, if those who carry guns can be treated in this fashion, so can all others whose work, if performed under the influence of drugs, may endanger others— automobile drivers, operators of other potentially dangerous equipment, construction workers, school crossing guards. A similarly broad scope attaches to the Court's approval of drug testing for those with access to "sensitive information." Since this category is not limited to Service employees with drug interdiction duties, nor to "sensitive information" specifically relating to drug traffic, today's holding apparently approves drug testing for all federal employees with security clearances—or, indeed, for all federal employees with valuable confidential information to impart. Since drug use is not a particular problem in the Customs Service, employees throughout the government are no less likely to violate the public trust by taking bribes to feed their drug habit, or by yielding to blackmail. Moreover, there is no reason why this super-protection against harms arising from drug use must be limited to public employees; a law requiring similar testing of private citizens who use dangerous instruments such as guns or cars, or who have access to classified information, would also be constitutional.

It Is Unjustifiable to Impair Liberties Just to Make a Point

There is only one apparent basis that sets the testing at issue here apart from all these other situations—but it is not a basis upon which the Court is willing to rely. I do not believe for a minute that the driving force behind these drug-testing rules was any of the feeble justifications put forward by counsel here and accepted by the Court. The only plausible explanation, in my view, is what the Commissioner himself offered in the concluding sentence of his memorandum to Customs Service employees announcing the program: "Implementation of the drug screening program would set an important example in our country's struggle with this most serious threat to our national health and security." Or as respondent's brief to this Court asserted: "[I]f a law enforcement agency and its employees do not take the law seriously, neither will the public on which the agency's effectiveness depends."

What better way to show that the Government is serious about its "war on drugs" than to subject its employees on the frontline of that war to this invasion of their privacy and affront to their dignity? To be sure, there is only a slight chance that it will prevent some serious public harm resulting from Service employee drug use, but it will show to the world that the Service is "clean," and—most important of all—will demonstrate the determination of the Government to eliminate this scourge of our society! I think it obvious that this justification is unacceptable; that the impairment of individual liberties cannot be the means of making a point; that symbolism, even symbolism for so worthy a cause as the abolition of unlawful drugs, cannot validate an otherwise unreasonable search.

There is irony in the Government's citation, in support of its position, of Justice [Louis] Brandeis' statement in *Olmstead v. United States* (1928), that "[f]or good or for ill, [our Government] teaches the whole people by its example." Brandeis was there *dissenting* from the Court's admission of evidence obtained through an unlawful Government wiretap. He was not praising the Government's example of vigor and enthusiasm in combatting crime, but condemning its example that "the end justifies the means." An even more apt quotation from that famous Brandeis dissent would have been the following:

> [I]t is . . . immaterial that the intrusion was in aid of law enforcement. Experience should teach us to be most on our guard to protect liberty when the Government's purposes are beneficent. Men born to freedom are naturally alert to repel invasion of their liberty by evil-minded rulers. The greatest dangers to liberty lurk in insidious encroachment by men of zeal, well-meaning, but without understanding.

Those who lose because of the lack of understanding that begot the present exercise in symbolism are not just the Customs Service employees, whose dignity is thus offended, but all of us—who suffer a coarsening of our national manners that ultimately give the Fourth Amendment its content, and who become subject to the administration of federal officials whose respect for our privacy can hardly be greater than the small respect they have been taught to have for their own.

> "We like the easy way out, and mandatory drug screening seems cheap; it only costs us our Fourth Amendment security."

Mandatory Drug Screening Weakens the Protection of the Fourth Amendment

George J. Annas

George J. Annas is a professor of health law and director of the Law, Medicine and Ethics Program at Boston University Schools of Medicine and Public Health. In the following viewpoint, he points out that because of two 1989 Supreme Court decisions, the Fourth Amendment offers citizens far less protection against governmental searches than in the past and that this will "inevitably result in a lessening of respect for the bodily integrity of all citizens." He argues that this trend may lead not only to expansion of testing for illegal drugs, but to intrusive screening tests for many kinds of health conditions and even to genetic screening for susceptibility to disease. In his opinion, this is part of a growing tendency to focus on individual citizens' shortcomings instead of on social problems for which solutions should be sought.

Now that the use of illegal drugs, especially crack, has been labeled the country's number one public health and safety issue, we are likely to see more and more creative and repressive ways to attack the problem. As long as the use of such drugs and its associated violence was largely confined to

George J. Annas, "Crack, Symbolism, and the Constitution," *Hastings Center Report*, vol. 19, May/June 1989, pp. 35–37. Copyright © 1989 Hastings Center. Reproduced by permission.

America's underclass and inner cities, their use was not seen as a central societal issue. But times seem to have changed; what was commonplace drug use in the 1960s, for example, is now grounds for disqualification for positions in law enforcement, and even for the post of Justice of the Supreme Court of the United States. Much of the recent reaction to drug-taking can be viewed as sanctimonious, hysterical, and hypocritical.

On the other hand, both legal and illegal drugs can impair job performance, and employers have legitimate concerns about employee drug use in particular settings. How far should employers be able to go in testing and screening their employees? The Supreme Court decided its first two cases on this question in March [1989]. Although the rulings leave much to be decided and debated at a later date, they begin to set the limits of governmentally mandated drug testing.

The Railroad Rules

The first case [*Skinner v. Railway Labor Executives' Association*] involved regulations promulgated by the Secretary of Transportation under the Federal Railroad Safety Act of 1970. These regulations, announced in 1985, were based in part on a 1979 study that concluded that 23 percent of railroad operating personnel were "problem drinkers," and statistics that showed that from 1975 through 1983 there were forty-five train accidents involving "errors of alcohol and drug-impaired employees" resulting in thirty-four deaths and sixty-six injuries. The regulations required the collection of blood and urine samples for toxicological testing of railroad employees following any accident resulting in a fatality, release of hazardous materials or railroad property damage of $500,000; and any collision resulting in a reportable injury or damage to railroad property of $50,000. After such an event, all crew members are to be sent to an independent medical facility where both blood and urine samples will be obtained from them in an attempt to determine the cause of the incident. Employees who refuse testing may not perform their job for nine months. Under another provision, the railroad may perform breath and urine tests on individuals it has "reasonable suspicion" are under the influence of drugs or alcohol while on the job.

The Customs Service Rules

In 1986 the Commissioner of Customs announced the implementation of a drug-testing program for certain Customs officials, finding that although "Customs is largely drug-free . . . unfortunately no segment of society is immune from the threat of illegal drug use." Drug testing was made a condition of placement or employment for positions that meet one of three criteria: They have direct involvement in drug interdiction or enforcement of drug laws; require the carrying of firearms; or require handling classified material. After an employee qualifies for a position covered by the rule, he or she is notified by letter that final selection is contingent upon successful completion of drug screening. An independent drug-testing company contacts the employee and makes arrangements for the urine test. The employee is required to remove outer garments and personal belongings, but may produce the sample behind a partition or in the privacy of a bathroom stall. A monitor of the same sex, however, is required to remain close at hand "to listen for the normal sounds of urination," and dye is added to the toilet water to prevent adulteration of the sample. The sample is then tested for the presence of marijuana, cocaine, opiates, amphetamines, and phencyclidine. Confirmed positive results are transmitted to the medical review officer of the agency, and can result in dismissal from the Customs Service. Test results may not, however, be turned over to any other agency, including criminal prosecutors, without the employees' written consent.

The Fourth Amendment's Protection

The Fourth Amendment provides: The right of the people to be secure in their persons, houses, papers, and effects, against unreasonable searches and seizures, shall not be violated, and no Warrants shall issue, but upon probable cause, supported by Oath or affirmation, and particularly describing the place to be searched, and the person or things to be seized.

Constitutional protections, of course, apply only to acts of the government (and not to those of private employers). The Court, however, had no trouble concluding . . . that the railroad was acting as an instrument of the government. Of course, tests carried out on Customs Service employees, done pursuant to government rule, are covered by the Constitution.

The Court, in opinions written by Justice Anthony Kennedy, easily found that blood and urine tests are "searches" under the Fourth Amendment. As to the taking of a blood sample, the court noted, "this physical intrusion, penetrating beneath the skin, infringes an expectation of privacy that society is prepared to recognize as reasonable." Moreover, the chemical analysis of the blood "is a further invasion of the tested employee's privacy interests." Breath tests, the court found, implicated "similar concerns about bodily integrity and . . . should also be deemed a search." And although urine tests do not entail a surgical intrusion, "analysis of urine, like that of blood, can reveal a host of private medical facts about an employee, including whether she is epileptic, pregnant or diabetic." The collection of urine also intrudes upon the reasonable expectation of privacy, especially when accompanied by a visual or aural monitor, and the Court amazingly and perhaps prudishly seems to think that having one's urination monitored is a greater violation of privacy than having a needle penetrate one's skin to remove blood.

A "Reasonable" Search?

Is this the type of search that requires a warrant, and if not, is this type of search "reasonable?" The Court concluded that no

warrant is required for two primary reasons: (1) there are virtually no facts for a neutral magistrate to evaluate (because
both the circumstances justifying the search and the limits of
the search are "defined narrowly and specified in the regulations that authorized them"); and (2) the delay needed to procure a warrant might result in the destruction of valuable evidence as the drugs and/or alcohol metabolize in the employee.
Warrantless searches have traditionally required at least some
showing of "individualized suspicion"; but the court made it
clear that "a search may be reasonable despite the absence of
such suspicion" at least if the interference is "minimal" and it
takes place in the employment context. This is because "[o]r-
dinarily, an employee consents to significant restrictions on
his freedom of movement where necessary for his employment, and few are free to come and go as they please during
working hours."

Blood tests are not unreasonable because they are usually
"taken by a physician . . . according to accepted medical practice" and are "safe, painless, and commonplace." Breath tests,
although not done by medical personnel, involve no piercing
of the skin and can be done "with a minimum of inconve-
nience or embarrassment." Urine testing was seen as the most
difficult to justify, because it involves "an excretory function
traditionally shielded by great privacy." However, the railroad
regulations do not require direct observation, and the test is
done in a medical environment "by personnel unrelated to the
railroad employer, and is thus not unlike similar procedures
encountered often in the context of a regular physical examination."

More important, however, seems to have been the Court's
view of the "diminished" expectation of privacy employees
have when they enter certain occupations. The railroad industry was described as "regulated pervasively to ensure
safety. . . ." The court notes that an "idle locomotive" is harm-
less, but "it becomes lethal when operated negligently by per-

sons who are under the influence of alcohol or drugs." Customs employees are involved in drug interdiction and so "reasonably should expect effective inquiry into their fitness and probity." This is especially true, the court opined, because "drug abuse is one of the most serious problems confronting society today [and] the almost unique mission of the Service gives the Government a compelling interest in ensuring that many of these covered employees do not use drugs even off duty, for such use creates risks of bribery and blackmail against which the Government is entitled to guard."

Unlike the case with railroad employees, however, there was virtually no evidence that drug use is a problem in the Customs Service. [In *National Treasury Employees v. Von Raab*], only five of 3,600 screened employees tested positive for drug use. The court, however, termed this finding a "mere circumstance," concluding that the Customs program was designed to prevent the promotion of drug users to sensitive positions, as well as to detect such employees. In the Court's words:

> The Government's compelling interests in preventing the promotion of drug users to positions where they might endanger the integrity of our Nation's borders or the life of the citizenry outweigh the privacy interests of those who seek promotion to these positions, who enjoy a diminished expectation of privacy by virtue of the special, and obvious, physical and ethical demands of those positions.

Nonetheless, the court sent back to the lower court the section of the rules that dealt with individuals who had access to "classified" material for further proceedings because the list of positions covered (which ran from attorney to messenger) seemed overly broad to meet the goals of the rule.

The Dissents

The railroad decision was a 7 to 2 opinion, with Justice Thurgood Marshall dissenting. Justice Marshall saw the opinion as gutting the Fourth Amendment. He agreed that "declaring a war on illegal drugs is good public policy," but concluded that "the first, and worst, casualty of the war will be the precious liberties of our citizens." He was especially critical of the Court's reading into the Fourth Amendment an exception to the probable cause requirement when "special needs" of non-law-enforcement agencies make either warrants or probable cause inconvenient requirements. The Court's reasoning, he noted, equated past cases that dealt with minimal searches of a person's possessions, with a search of the person's body itself, and thereby widened the "special needs" exception without requiring any evidence of wrongdoing on the part of the person.

The most powerful and surprising dissenting voice, however, was raised by Justice Antonin Scalia in the 5 to 4 Customs case. Like the railroad case, the Customs case was driven by the "war on drugs" and "the Government's compelling interests in safety and in the integrity of our borders." As Justice Scalia argued, however, whereas there was evidence of drug and alcohol abuse resulting in accidents and injury in the railroad case, in the Customs case "neither frequency of use nor connection to harm is demonstrated or even likely." He concluded therefore that the "Customs Service rules are a kind of immolation of privacy and human dignity in symbolic opposition to drug use." Justice Scalia noted, for example, that the record discloses not even one incident of the speculated harms actually occurring, nor even one incident "in which the cause of bribetaking, or of poor aim, or of unsympathetic law enforcement . . . was drug use." Instead, the evidence is that the Customs Service is "largely drug free." The regulations thus expose "vast numbers of public employees" to the "needless indignity" of urine screening simply so the Customs Service can "set an important example" in the country's fight against drugs. Justice Scalia properly concluded that such a solely symbolic justification is "unacceptable" when the Fourth Amendment is violated:

... the impairment of individual liberties cannot be the means of making a point; ... symbolism, even symbolism for so worthy a cause as the abolition of unlawful drugs, cannot validate an otherwise unreasonable search.... Those who lose ... are not just the Customs Service employees, whose dignity is thus offended, but all of us—who suffer a coarsening of our national manners that ultimately give the Fourth Amendment its content, and who become subject to the administration of federal officials whose respect for our privacy can hardly be greater than the small respect they have been taught to have for their own.

The Age of Screening

Justices Marshall and Scalia both seem correct. The Fourth Amendment now affords citizens far less protection against governmental searches, and this will inevitably result in a lessening of respect for the bodily integrity of all citizens.

As Justice Scalia quite correctly noted, the opinions can immediately be extended to all those employees who carry firearms, and from there, to all employees whose jobs, if performed under the influence of drugs, could harm the public, including: "automobile drivers, operators of potentially dangerous equipment, construction workers, school crossing guards."

The war on drugs, of course, has special application to physicians. They, as much as Customs officials, might be chosen by the government as a "model" for other citizens. Since they have direct access to drugs and write prescriptions for them, the government could require drug screening as a condition of licensure. And, if it could be shown that physicians, nurses, and other health care professionals have a drug or alcohol problem that could seriously harm citizens, this could permit states and the federal government to pass laws requiring drug screening for employment or the granting of staff privileges. Of course, private employers can adopt such rules themselves now, and are not subject to the provisions of the

Fourth Amendment. On the other hand, some state constitutions are more protective than the federal constitution, and in those states, state and local governmental employees may be protected from mandatory screening in the absence of probable cause.

These rulings leave open a number of questions of relevance to the forty-four cases currently [in 1989] in the federal courts dealing with various aspects of drug screening tests. Perhaps the most important is whether the government can require random drug screening, not related to a specific incident, reason to suspect drug use, hiring or promotion. We can only speculate, but it does seem reasonable to conclude that Justice Kennedy would find nothing constitutionally infirm in a practice of random drug screening to further the "war on drugs." The four dissenters in the Customs case would surely rule against random drug screening, at least in the absence of a specific finding that the class of employees screened had a history of drug use that could result in serious injuries.

The cases also raise a generic issue about screening. If the government can compel screening to determine if an individual is using a substance whose use is forbidden by the criminal law, is there anything the government cannot mandate screening for to protect the public? For example, what would prevent the government from requiring genetic screening for susceptibility to various diseases if and when such screening becomes available, and assuming the screened-for disease rendered the individual "less qualified" for the job, or potentially dangerous on the job? Indeed, the Court seems almost eager to view any screening test done by a physician as routine medical practice as de facto reasonable, without any real analysis of how intrusive such testing can be.

Will the courts support mandatory screening only for rank-and-file and "blue collar" workers, and balk at any mandatory attempt to screen professionals, such as lawyers, physicians, professors, and high-ranking government officials? How the court ultimately rules on mandatory screening for those with access to "classified" documents will tell us much about the Court's view of which "class" of citizen deserves to be screened, and which class deserves the protection of the Fourth Amendment.

As the "age of screening" dawns, there will be more airport searches, more highway road blocks for sobriety testing, more intense border searches, and more screening tests for hiring, promotion, licensing, and continued employment. It is all part of our penchant to focus on individual citizens instead of systemic social problems. We seem to think that by treating citizens like grapes from Chile to be screened for cyanide, or apples to be screened for Alar we can solve our problems of poverty, racism, and violence. It is the impulse that had us concentrate on the captain's drinking when the Exxon Valdez ran aground and spilt its environmentally lethal cargo in Alaska's Prince William Sound. Instead of reexamining the activities and planning of the oil industry to prevent and contain oil spills, we will likely see more drug and alcohol screening of tanker crews as the solution to oil spills. We like the easy way out, and mandatory drug screening seems cheap; it only costs us our Fourth Amendment security.

> "Government action [is] symbolic when
> it is not likely to remedy an existing
> harm, or if indeed the existence of the
> harm to be remedied has not been es-
> tablished."

Court Justifications for Rulings in Drug Cases Are Often Largely Symbolic

Douglas Husak and Stanton Peele

Douglas Husak is a professor of philosophy and law at Rutgers University. Stanton Peele is a psychologist who has done extensive research on addiction and has written many controversial books and articles about it. In the following viewpoint, the authors exam Skinner v. Railway Labor Executives' Association *and* National Treasury Employees Union v. Von Raab, *two 1989 cases in which the Supreme Court justified government action that overrides constitutional rights on the basis of the harm done by drugs. The authors argue that the action was merely symbolic—that it was not likely to prevent harm and, in the cases involved, there was no evidence that any significant danger of harm existed.* Skinner *dealt with the testing of railroad employees, while* Von Raab *dealt with testing of Customs Service personnel. Husak and Peele contend that in neither case was there any evidence of a drug problem. The problem among railroad workers had been alcohol abuse, not drug use, and use of illegal drugs among Customs workers was extremely rare. Moreover, no evidence of drug testing's effectiveness was considered.*

Douglas Husak and Stanton Peele, "'One of the Major Problems of Our Society': Symbolism and Evidence of Drug Harms in U.S. Supreme Court Decisions," *Contemporary Drug Problems*, Summer 1998, pp. 191–233. Copyright © 1996-2008 Stanton Peele. All rights reserved. Reproduced by permission.

The Court based its decisions simply on harm that might occur, because, say the authors, the justices shared the public's heightened fear that the danger to society from drugs was exceptionally serious.

When a government action impinges on important individual rights, the United States Supreme Court is compelled to identify the harms it perceives in a behavior or practice in order to assess the constitutionality of a statute, a government policy, or a penalty. We select[ed] the six most recent cases in which the Court has expressed views about the harms it attributes to drug use. We conclude that the recent history of such Court justifications shows that they are almost entirely symbolic. Although several Justices in minority opinions have themselves made this point, only in [*Chandler v. Miller* (1997)] did the majority label a government action against drugs as symbolic and thereby conclude that it did not justify the infringement of constitutional rights involved. . . .

For the purposes of this article, we define government action as symbolic when it is not likely to remedy an existing harm, or if indeed the existence of the harm to be remedied has not been established. As we will see, the allegation that a drug proscription, policy or penalty is "merely symbolic" has frequently been made by Justices who believed a law to be unconstitutional. . . .

Background to Court Decision-Making: Specifying Drug Harms

The efficacy of U.S. policies against illicit drugs—including the legitimacy of criminal prohibitions and sanctions—is the subject of enormous controversy. Much of this controversy stems from wide disagreement about the exact nature of the harms that drug legislation is designed to prevent, and whether these harms can effectively be prevented by the legislation in question. What are these harms? How persuasive is the evidence that they really exist? How great is the likelihood that the law under scrutiny will help to prevent them? How important is the state interest in curtailing these harms, especially when the measures employed to prevent them infringe upon constitutional rights?

Although legislators have no general obligation to describe the rationale for the prohibitions and penalties they enact, in some instances courts may be required to identify the harms they believe that legislation is designed to prevent and to assess the prospects that the law in question will be effective in preventing them. Of course, courts are divided about these matters as well. Occasionally courts have vehemently opposed the drug regulations they have reviewed. When constitutional issues are raised, the Supreme Court is the final arbiter in such cases. The Court has never squarely addressed the justification of drug prohibitions. In a handful of exceptional cases, however, the Court has presented, often tangentially, what it understands the rationale for various drug laws to be. These occasions arise when a drug offense, penalty, or preventive measure is alleged to infringe on a constitutional right. . . .

All in all, the Supreme Court has analyzed the harms of drug use in only a very few cases. In some of these cases, the compelling state interest test was applied because a drug-prohibition-related activity burdened (at least in the eyes of some of the justices) a fundamental right clearly embodied in the Constitution. Sometimes, when a rational basis test rather than a compelling state interest test was invoked, the Court has nonetheless pronounced upon the harms of drug use. In these cases, the Court has balanced the infringement upon individual freedoms against the "reasonableness" of the drug regulation. A primary type of case that often (but not always) requires the Court to identify the state's purposes underlying an action occurs when the government conducts a search for drugs. The Court is then seemingly obligated to evaluate whether the state's goals are important enough to outweigh the individual's Fourth Amendment protection. . . .

One would hope that the infringement upon constitutional rights would require more than a broad allegation that "the use of illegal drugs . . . is one of the major problems of our society." Even if this allegation is true, one would expect that the Court would require evidence that specific harms to be prevented actually exist and that the legislation in question is reasonably effective in preventing them. . . .

In examining [recent] cases, we could detect no authoritative declaration of a rationale for proscribing the use of illicit drugs generally, or of any given drug in particular—or even an evidentiary basis for making the decision. We find, rather, that these cases provide an insight into popular prejudices about drugs, in addition to the idiosyncratic thinking of various Supreme Court Justices about the harms of drugs. Both the nature and the severity of the harms alleged are open to question. Indeed, these allegations are frequently challenged in dissenting opinions. . . .

Skinner v. Railway Labor Executives' Association

Skinner [*v. Railway Labor Executives' Association*] is the first of the cases decided by the Supreme Court in the last decade to have addressed the rationale for drug policies. At issue was the right of the Federal Railroad Administration (FRA) to test railroad employees for drugs and alcohol upon the occurrence of accidents or other untoward events. . . .

In the 1970s, the efforts to curb alcohol use were broadened to include illicit drugs. But no evidence of drug problems among railway workers was cited to support the expansion of these deterrent efforts. Instead, the data used to substantiate the need for testing in *Skinner* did not differentiate between the harms caused by alcohol and those attributed to drugs. . . .

Although *Skinner* included only the most generic statements about drug use, the case became the keynote in justifying future Supreme Court decisions in drug cases. Consider, for example, the citation to *Skinner* from [*National Treasury Employees Union*] *Von Raab*, announced by the Supreme Court on the same day: "Petitioners . . . do not dispute, nor can there be doubt, that drug abuse is one of the most serious problems confronting our society today. There is little reason to believe that American workplaces are immune from this pervasive social problem, as is amply illustrated by our decision in [*Skinner v.*] *Railway Labor Executives.*" No data were cited in *Skinner* to establish the levels of drug use in the population to be tested, let alone throughout society. Had the incidence of drug abuse really been so "pervasive," hard evidence of its existence should have been easy to supply. The primary question presented in *Skinner* is how a case involving no evidence of drug use came to stand as authority for the proposition that drug abuse is an obvious problem in the American workplace and throughout society at large. . . .

The claim that the state has a compelling interest in the prevention of railway accidents cannot be contested. And the claim that impairment caused by drugs or alcohol might contribute to such accidents seems fairly incontrovertible. Here, then, is an obvious harm, and a plausible allegation about how drug use might cause it. Still, the broad statements in *Skinner* do not comprise concrete evidence of a drug problem that can be readily employed in balancing the need for intrusive testing. Rather, the unanimous fear of drug abuse in society was asserted and accepted uncritically; the concurring opinion of [Justice John Paul] Stevens and even the dissenting opinion by [Justice Thurgood] Marshall and [Justice William J.] Brennan did not question whether this fear was exaggerated or misplaced. . . .

National Treasury Employees Union v. Von Raab

In a companion case to *Skinner*, a union of federal employees brought suit against the United States Customs Service to challenge the constitutionality of its drug-testing program. The program required urine specimens from employees who were required to handle classified materials, to inspect for illegal drugs, or to carry firearms. Employees who tested positive were subject to dismissal from the service. . . .

Judging the reasonableness (and thus the constitutionality) of the drug-testing program requires a balancing of "the individual's privacy expectations against the Government's interests." To perform this balancing, the Court in *Von Raab* (as with *Skinner*) had to identify and quantify the importance of deterring drug use in a specific population. While *Skinner* presented survey evidence about the use of alcohol, generalized from such data to drug use, and then made an inference about the problems this use could be expected to produce among railway workers, *Von Raab* presented no evidence of any kind of drug use among the relevant population. Instead, the Court simply found that the state has a "special need" to deter any drug use that *might* occur among Customs employees in sensitive positions.

The Court's analysis described the government's interest in deterring drug use and the weight it should be given when balanced against Fourth Amendment values. The results of this analysis varied with the particular position of the employee in question. The menace of Customs employees who are authorized to use deadly force but are impaired by drugs seems somewhat analogous to that of railway employees who are impaired. These employees "discharge duties fraught with such risks of injury to others that even a momentary lapse of attention can have disastrous consequences." Thus the Court found "that the public should not bear the risk that employees who may suffer from impaired perception and judgment will be promoted to positions where they may need to employ deadly force. Indeed, ensuring against the creation of this dangerous risk will itself further Fourth Amendment values, as the use of deadly force may violate the Fourth Amendment in certain circumstances."

The Court extended this reasoning to find that interests of comparable weight support the testing of employees who simply inspect for illicit drugs. The Supreme Court quoted with approval the speculation by the Court of Appeals that drug use among this category of employees would cast "substantial doubt on their ability to discharge their duties honestly and vigorously, undermining public confidence in the integrity of the Service and concomitantly impairing the Service's efforts to enforce the drug laws." Moreover, the Court feared that illicit drug users in this population "may be tempted to divert for their own use portions of any drug shipments they interdict" and be "susceptible to bribery and blackmail." The Court concluded that the "national interest in self-protection could be irreparably damaged if those charged with safeguarding it were, because of their own drug use, unsympathetic to their mission of interdicting narcotics." Finally, the Court found that the testing of employees who handled classified documents did not satisfy the test of reasonableness, and remanded this part of the decision to the Court of Appeals for further clarification of its scope and purpose.

The petitioning customs workers argued that the drug-testing program would not deter drug use, since employees were notified five days in advance of a pending test and could avoid a positive urine by abstaining from drug use for that brief period. Fundamentally, however, the union argued that drug testing was unjustified because the Customs Service had failed to provide any evidence of a drug problem in the group: "Counsel for petitioners informed us at oral argument that no more than five employees out of 3,600 have tested positive for drugs." The Court's response to this argument is noteworthy: "Petitioners do not dispute, nor can there be doubt, that drug abuse is one of the most serious problems confronting our society today. . . . The mere circumstance that all but a few of the employees tested are entirely innocent of wrongdoing does not impugn the program's validity." . . .

[Justice Antonin] Scalia's dissent, however, was highly critical of the Court's use of evidence in *Von Raab* in balancing individual rights against government needs. Joined by [Justice John Paul] Stevens, Scalia protested that "in the present case . . . neither frequency of use nor connection to harm is demonstrated or even likely. In my view, the Customs Service rules are a kind of immolation of privacy and human dignity in symbolic opposition to drug use. . . . [T]he Court's opinion in the present case . . . will be searched in vain for real evidence of a real problem that will be solved by urine testing of Customs Service employees." . . .

Scalia regarded the state's alleged justifications for implementing drug-testing programs as so "feeble" that he was led to speculate about the real motivation behind these initiatives: the tests must be designed to demonstrate "that the Government is serious about its 'war on drugs.'" In language that would become familiar in dissenting opinions involving drugs until *Chandler*, Scalia warned that "symbolism, even symbolism for so worthy a cause as the abolition of unlawful drugs, cannot validate an otherwise unreasonable search."

In *Von Raab*, four justices would not allow the infringement upon Fourth Amendment protections to prevent a problem that might have been entirely imaginary. For the majority, even hypothetical drug use was sufficient to decide that a search of body fluids is reasonable. Of course, the evidence of specific drug use in *Skinner*, in which Scalia and Stevens joined the majority, was little better than that in *Von Raab*. Nonetheless, Scalia and Stevens distinguished *Skinner* by the "demonstrated frequency of drug and alcohol use by the target class of employees, and the demonstrated connection between such use and grave harm." In making this argument, Scalia (and Stevens) did not differentiate between alcohol use and drug use. Scalia might have been on firmer ground if he had contrasted the importance of the state's interest in deterring drug use between the two target populations. The specter of a drugged railway engineer who negligently endangers the lives of passengers seems more worrisome than that of an agent involved in drug interdiction whose own drug use compromises his mission.

Moreover, the tests in *Skinner* were imposed only after a railroad accident had actually occurred. In contrast, the tests in *Von Raab* were imposed as a proactive measure, to reduce the risk of *a possible* harm. The majority defended the importance of drug-testing Customs workers merely by asserting the significance of the workers' mission: "The Customs Service is our Nation's first line of defense against one of the greatest problems affecting the health and welfare of our population." And "there is little reason to believe that American workplaces are immune from this pervasive social problem, as is amply illustrated by our decision in [*Skinner v.*] *Railway Labor Executives*." Rather than demanding evidence of drug harms, the Court simply bootstrapped its decision onto an equally unsubstantiated claim in an earlier case. . . .

The Absence of Evidence

Perhaps most significantly, each of [the cases we examined] authorized the state to take action in the absence of clear evidence that any identified drug harm had actually occurred. Small wonder the critics of these judgments believed the state interest in combating drugs to be "merely symbolic."

These cases are diverse. Taken together, they support the following six generalizations:

1. *Each case reflects the heightened concern over drugs that began in the 1980s.* Drug use became an overwhelming concern for American society late in the 1980s. For a period of sixteen months from 1989 through 1990, a majority of Americans identified drugs as the country's number one social problem. This impelled anti-drug initiatives throughout the government that invariably were supported by the public. The Supreme Court re-

flected this concern and activity. *Skinner*, for example, dealt with railroad workers, for whom alcohol was clearly the primary object of abuse. With the advent of the drug scare of the 1980s, however, reasonable concerns about safety and impairment became associated with drugs and were used to justify random urine testing. Drug use among adults (and presumably among railway workers) was declining before the tests were implemented, including in many settings where testing did not occur. The FRA's decision to require these tests, and the Court's approval of them, are best explained by the heightened concern about drug use in society generally.

2. *Evidence of actual drug use in the populations affected was often highly indirect or even nonexistent.* Although none of these cases includes clear-cut evidence of the occurrence of the various harms said to be caused by drug use, the absence of evidence of actual drug use was most apparent in *Von Raab*. No data were presented to establish that Customs workers took drugs or that the alleged harms of drug use (e.g., blackmail, susceptibility to payoffs, or misuse of firearms) had ever taken place. The government's rationale for drug testing, which the Court accepted as compelling, was expressed wholly in the form of hypotheticals. . . .

3. *The American drug scourge was supported in subsequent cases by unsubstantiated Court precedents in earlier cases.* The Court's allegation in *Skinner* that drugs are a grave social menace was regularly invoked to bootstrap this conclusion into later decisions despite the lack of supporting data. *Von Raab* referred to this conclusion in *Skinner* and then itself became the touchstone for all further decisions linked to a proposed national drug scourge. The original basis for this conclusion, however, is obscure and unsubstantiated.

4. *Citations of data about drug use and about the effective-ness of proposed measures to deter it are selective and un-critical.* The justices made little reference to data on the incidence of drug use—none to national drug use statis-tics, almost none to drug use by the populations in-volved in the cases under discussion, and none to the epidemiology of drug abuse/addiction and its relation to drug use. Moreover, almost no evidence was evaluated about the efficacy of drug-testing programs or of any other remedy allowed in any of these cases. On the few occasions when the justices cited social science studies—for example, research on adolescent drug use/abuse and on the connection between drugs and crime—their in-terpretations of the data are dubious.

5. *Critical stances about drug policies were adopted in dis-senting opinions.* In one case or another, several jus-tices—Marshall, Brennan, Stevens, Scalia, [Harry] Black-mun, [Byron] White, [Sandra Day] O'Connor and [David] Souter—expressed serious reservations about the challenged drug proscription/penalty/test. . . .

6. *Acceptance of the drug menace was unanimous among Supreme Court Justices.* Despite the skepticism expressed in these dissenting opinions, no justice emerges as a consistent critic of America's drug policies. Even those justices who might have been thought most likely to resist anti-drug measures—such as Marshall and Bren-nan—were inclined to base their reservations on grounds that seem more procedural than substantive. Marshall, like other justices, typically prefaced his objec-tions by indicating his belief in the dire and imminent dangers of drug use. With the retirements of Marshall, Brennan and Blackmun, no justice seems inclined to adopt a consistently skeptical position toward the legiti-macy of government actions against drug users.

CHAPTER 2

Civil Forfeiture and Criminal Punishment for the Same Offense Is Legal

Case Overview

United States v. Ursery (1996)

In 1992, the Michigan police got a warrant to search Guy Ursery's house after finding marijuana growing on what they mistakenly thought was his property. Inside they found marijuana seeds, stalks and a grow light. They did not claim that he had sold any marijuana; it had been used only by him and his family. Nevertheless, they began civil forfeiture proceedings against his house on the grounds that it had been used for the unlawful processing of a controlled substance. This meant that although Ursery had not yet been convicted of a crime, his home would be forfeited to the government.

Many people are surprised to learn that the government can seize a home or other property before bringing a charge against its owner. However, this is commonly done. Sometimes the property is not returned even if the owner is found to be blameless. Although since passage of the Civil Asset Forfeiture Reform Act of 2000 (CAFRA) assets cannot be taken from innocent parties by the federal government, the owner is presumed guilty unless proven innocent, rather than the other way around as in criminal trials. The constitutional rights that normally protect crime suspects do not apply to civil forfeitures. This is because in civil forfeiture cases, which in legal terminology are called *in rem* proceedings, the action is taken against the property rather than against a person. Cases of this kind have titles such as *One 1958 Plymouth Sedan v. Pennsylvania, United States v. $160,000 in U.S. Currency*, and *United States v. One Assortment of 89 Firearms*, all of which are real cases. Legally, the property itself is considered guilty of wrongdoing; the owner need not even be aware that forfeiture action is being taken.

Civil forfeiture has been used since ancient times and the first U.S. law authorizing it was passed in 1789. However, it has been employed much more frequently since the war on drugs began. It is one of the most effective weapons against large-scale drug traffickers, who would otherwise find ways of hiding or disposing of their assets before being brought to trial. No one objects when the government confiscates contraband or takes the profit out of crime by seizing its proceeds. However, many people believe that forfeiture laws are unfair when they are used to seize legally acquired property merely because it was associated in some way with lawbreaking.

Guy Ursery's house had not been purchased with proceeds of unlawful activity. He had nevertheless grown marijuana there, so although forfeiture of his home seemed an excessive penalty, he had no chance of contesting it. He did not actually lose the house, as he was able to settle the claim against it by paying the government $13,250. But after that, he was arrested, convicted of manufacturing a controlled substance, and sentenced to sixty-three months in prison. He then appealed on the grounds that under the double jeopardy clause of the Fifth Amendment, the Constitution forbids punishing a person twice for the same offense. The Court of Appeals agreed with this argument and reversed his conviction, whereupon the government appealed to the U.S. Supreme Court.

In considering the case, the Supreme Court combined it with another case, that of Charles Wesley Arlt and James Wren, who were convicted of conspiracy to aid and abet the manufacture of methamphetamine and to launder money. The government had filed an *in rem* complaint against a bank account, a helicopter, a shrimp boat, an airplane, and numerous automobiles, which it alleged were connected to the crimes. It did not seize this property until after Arlt and Wren had been prosecuted, and they therefore argued that the forfeiture violated the double jeopardy clause.

The Supreme Court case hinged on the issue of whether civil forfeiture constitutes punishment. It might seem obvious that it does—but since under the law only the property itself is involved in such proceedings, technically the penalty is paid by the property. The Supreme Court, citing a number of precedents, ruled eight to one that for the purposes of double jeopardy, civil forfeiture, as distinguished from payment of a fine, is not intended as punishment of the property's owner. A person can therefore lose property and be sent to prison for the same act, even if—as in Ursery's case—the two proceedings are independent of each other and are separated by time.

This decision was applauded by federal prosecutors, who had been reluctant to use civil forfeiture in drug-trafficking cases lest that prevent criminal prosecution of the guilty parties. But many civil rights advocates were appalled. Robert Fogelnest, president of the National Association of Criminal Defense Lawyers, said, "To anyone concerned about the growth of government power beyond constitutional limits, today may be the single saddest day in many [Supreme Court] terms."

The Court's Decision: Both Civil Forfeiture and Criminal Penalties Can Be Imposed for the Same Offense

William Rehnquist

William Rehnquist became a justice of the U.S. Supreme Court in 1972, and in 1986, he became chief justice, a position he held until his death in 2005. He was a strong conservative who believed in a strict interpretation of the Constitution. In the opinion of the Court in United States v. Ursery *in 1996, Rehnquist states that the imposition of civil forfeiture and criminal penalties for the same offense is not double jeopardy under the Fifth Amendment because civil forfeiture does not constitute punishment. Civil forfeiture proceedings are directed against the property involved, not against its owner. This has always been true, he says, citing a number of past cases in which the Court held that forfeiture was independent of punishment. This case is no different, Chief Justice Rehnquist declares. The aim of laws requiring forfeiture of property in drug cases is to discourage people from allowing their property to be used in connection*

William Rehnquist, majority opinion, *United States v. Ursery*, U.S. Supreme Court, June 24, 1996.

with drugs and to make sure they do not profit from drug deal-
ing, not to punish offenders. Therefore, the double jeopardy
clause does not apply to civil forfeiture.

Michigan Police found marijuana growing adjacent to respondent Guy Ursery's house, and discovered marijuana
seeds, stems, stalks, and a grow light within the house. The
United States instituted civil forfeiture proceedings against the
house, alleging that the property was subject to forfeiture because it had been used for several years to facilitate the unlawful processing and distribution of a controlled substance. Ursery ultimately paid the United States $13,250 to settle the
forfeiture claim in full. Shortly before the settlement was con-
summated, Ursery was indicted for manufacturing marijuana,
in violation of [the law]. A jury found him guilty, and he was
sentenced to 63 months in prison.

The Court of Appeals for the Sixth Circuit by a divided
vote reversed Ursery's criminal conviction, holding that the
conviction violated the Double Jeopardy Clause of the Fifth
Amendment of the United States Constitution. Ursery, in the
court's view, had ... been "punished" in the forfeiture proceeding against his property, and could not be subsequently
criminally tried.

Following a jury trial, Charles Wesley Arlt and James Wren
were convicted of: conspiracy to aid and abet the manufacture
of methamphetamine, conspiracy to launder monetary instruments, and numerous counts of money laundering. The District Court sentenced Arlt to life in prison and a ten-year term
of supervised release, and imposed a fine of $250,000. Wren
was sentenced to life imprisonment and a 5-year term of supervised release.

Before the criminal trial had started, the United States had
filed a civil *in rem* [action against property itself] complaint
against various property seized from, or titled to, Arlt and
Wren, or Payback Mines, a corporation controlled by Arlt. The

complaint alleged that each piece of property was subject to forfeiture both under [the law], which provides that "[a]ny property . . . involved in a transaction or attempted transaction in violation of" (the money laundering statute) "is subject to forfeiture to the United States" . . . More than a year after the conclusion of the criminal trial, the District Court granted the Government's motion for summary judgment in the civil forfeiture proceeding.

Arlt and Wren appealed the decision in the forfeiture action, and the Court of Appeals for the Ninth Circuit reversed, holding that the forfeiture violated the Double Jeopardy Clause. . . .

In the decisions that we review, the Courts of Appeals held that the civil forfeitures constituted "punishment," making them subject to the prohibitions of the Double Jeopardy Clause. The Government challenges that characterization of the forfeitures, arguing that the courts were wrong to conclude that civil forfeitures are punitive for double jeopardy purposes.

The Issue of Double Jeopardy Is Not New

Since the earliest years of this Nation, Congress has authorized the Government to seek parallel *in rem* civil forfeiture actions and criminal prosecutions based upon the same underlying events. . . .

And, in a long line of cases, this Court has considered the application of the Double Jeopardy Clause to civil forfeitures, consistently concluding that the Clause does not apply to such actions because they do not impose punishment.

One of the first cases to consider the relationship between the Double Jeopardy Clause and civil forfeiture was *Various Items of Personal Property v. United States* (1931). In *Various Items*, the Waterloo Distilling Corporation had been ordered to forfeit a distillery, warehouse, and denaturing plant, on the ground that the corporation had conducted its distilling business in violation of federal law. The Government conceded that the corporation had been convicted of criminal violations prior to the initiation of the forfeiture proceeding, and admitted that the criminal conviction had been based upon "the transactions set forth . . . as a basis for the forfeiture." Considering the corporation's argument that the forfeiture action violated the Double Jeopardy Clause, this Court unanimously held that the Clause was inapplicable to civil forfeiture actions:

> [This] forfeiture proceeding . . . is *in rem*. It is the property, which is proceeded against, and, by resort to a legal fiction, held guilty and condemned as though it were conscious instead of inanimate and insentient. In a criminal prosecution, it is the wrongdoer in person who is proceeded against, convicted, and punished. *The forfeiture is no part of the punishment for the criminal offense. The provision of the Fifth Amendment to the Constitution in respect of double jeopardy does not apply* [citations omitted; emphasis added].

In reaching its conclusion, the Court drew a sharp distinction between *in rem* civil *forfeitures* and *in personam* civil *penalties* such as fines: though the latter could, in some circumstances, be punitive, the former could not. . . .

Following its decision in *Various Items*, the Court did not consider another double jeopardy case involving a civil forfeiture for forty years. Then, in *One Lot Emerald Cut Stones v. United States* (1972), the Court's brief opinion reaffirmed the rule of *Various Items*. In *Emerald Cut Stones*, after having been acquitted of smuggling jewels into the United States, the owner of the jewels intervened in a proceeding to forfeit them as contraband. We rejected the owner's double jeopardy challenge to the forfeiture, holding that, "[if] for no other reason, the forfeiture is not barred by the Double Jeopardy Clause of the Fifth Amendment, because it involves neither two criminal trials nor two criminal punishments."

Noting that the forfeiture provisions had been codified separately from parallel criminal provisions, the Court determined that the forfeiture clearly was "a civil sanction." The forfeitures were not criminal punishments, because they did not impose a second *in personam* [action against a person] penalty for the criminal defendant's wrongdoing.

In our most recent decision considering whether a civil forfeiture constitutes punishment under the Double Jeopardy Clause, we again affirmed the rule of *Various Items. United States v. One Assortment of 89 Firearms* (1984), the owner of the defendant weapons was acquitted of charges of dealing firearms without a license. The Government then brought a forfeiture action against the firearms, alleging that they were used or were intended to be used in violation of federal law. In another unanimous decision, we held that the forfeiture was not barred by the prior criminal proceeding. . . .

Our inquiry proceeded in two stages. In the first stage, we looked to Congress' intent, and concluded that "Congress designed forfeiture under [the law] as a remedial civil sanction." This conclusion was based upon several findings. First, noting that the forfeiture proceeding was *in rem*, we found it significant that "actions *in rem* have traditionally been viewed as civil proceedings, with jurisdiction dependent upon the seizure of a physical object." Second, we found that the forfeiture provision, because it reached both weapons used in violation of federal law and those "intended to be used" in such a manner, reached a broader range of conduct than its criminal analogue. Third, we concluded that the civil forfeiture "further[ed] broad remedial aims," including both "discouraging unregulated commerce in firearms," and "removing from circulation firearms that have been used or intended for use outside regulated channels of commerce."

In the second stage of our analysis, we looked to "'whether the statutory scheme was so punitive either in purpose or effect as to negate' Congress' intention to establish a civil remedial mechanism." Considering several factors that we had used previously in order to determine whether a civil proceeding was so punitive as to require application of the full panoply of constitutional protections required in a criminal trial, we found only one of those factors to be present in the forfeiture. By itself, however, the fact that the behavior proscribed by the forfeiture was already a crime proved insufficient to turn the forfeiture into a punishment subject to the Double Jeopardy Clause. . . .

Since at least *Various Items*, we have distinguished civil penalties such as fines from civil forfeiture proceedings that are *in rem*. While a "civil action to recover . . . penalties, is punitive in character," and much like a criminal prosecution in that "it is the wrongdoer in person who is proceeded against . . . and punished," in an *in rem* forfeiture proceeding, "it is the property which is proceeded against, and by resort to a legal fiction, held guilty and condemned." Thus, though, for Double Jeopardy purposes, we have never balanced the value of property forfeited in a particular case against the harm suffered by the Government in that case, we have balanced the size of a particular civil penalty against the Government's harm. . . .

In the Cases Considered Here Forfeiture Is Not Punishment

We find that there is little evidence, much less the "'clearest proof'" that we require, suggesting that forfeiture proceedings, are so punitive in form and effect as to render them criminal despite Congress' intent to the contrary. The statutes involved in this case are, in most significant respects, indistinguishable from those reviewed, and held not to be punitive, in *Various Items*, *Emerald Cut Stones*, and *89 Firearms*.

Most significant is that [these statutes], while perhaps having certain punitive aspects, serve important nonpunitive goals. [The statute] under which Ursery's property was forfeited, provides for the forfeiture of "all real property . . . which is used or intended to be used, in any manner or part, to commit, or to facilitate the commission of" a federal drug felony. Requiring the forfeiture of property used to commit federal narcotics violations encourages property owners to take care in managing their property and ensures that they will not permit that property to be used for illegal purposes. . . .

The forfeiture of the property claimed by Arlt and Wren took place [under a statute that] provides for the forfeiture of "any property" involved in illegal money laundering transactions. [And one that] provides for the forfeiture of "[a]ll . . . things of value furnished or intended to be furnished by any person in exchange for" illegal drugs; "all proceeds traceable to such an exchange"; and "all moneys, negotiable instruments, and securities used or intended to be used to facilitate" a federal drug felony. The same remedial purposes served by [the above statute] are served by [these]. Only one point merits separate discussion. To the extent that [one] applies to "proceeds" of illegal drug activity, it serves the additional nonpunitive goal of ensuring that persons do not profit from their illegal acts.

Other considerations that we have found relevant to the question whether a proceeding is criminal also tend to support a conclusion that [these statutes] are civil proceedings. First, in light of our decisions in *Various Items, Emerald Cut Stones*, and *89 Firearms*, and the long tradition of federal statutes providing for a forfeiture proceeding following a criminal prosecution, it is absolutely clear that *in rem* civil forfeiture has not historically been regarded as punishment, as we have understood that term under the Double Jeopardy Clause. Second, there is no requirement in the statutes that we currently review that the Government demonstrates *scienter* [knowledge of wrongdoing] in order to establish that the property is subject to forfeiture; indeed the property may be subject to forfeiture even if no party files a claim to it and the Government never shows any connection between the property and a particular person. Though both contain an "innocent owner" exception, we do not think that such a provision, without more indication of an intent to punish, is relevant to the question whether a statute is punitive under the Double Jeopardy Clause. Third, though both statutes may fairly be said to serve the purpose of deterrence, we long have held that this purpose may serve civil as well as criminal goals. We recently [1996] reaffirmed this conclusion in *Bennis v. Michigan* where we held that "forfeiture ... serves a deterrent purpose distinct from any punitive purpose." Finally, though both statutes are tied to criminal activity, as was the case in *89 Firearms*, this fact is insufficient to render the statutes punitive. It is well settled that "Congress may impose both a criminal and a civil sanction in respect to the same act or omission," *Helvering* [*v. Mitchell* (1938)]. By itself, the fact that a forfeiture statute has some connection to a criminal violation is far from the "clearest proof" necessary to show that a proceeding is criminal.

We hold that these *in rem* civil forfeitures are neither "punishment" nor criminal for purposes of the Double Jeopardy Clause.

> *"There is simply no rational basis for characterizing the seizure of this respondent's home as anything other than punishment for his crime."*

Concurring and Dissenting Opinion: It Is Wrong to Impose Civil Forfeiture of Minor Offenders' Homes

John Paul Stevens

John Paul Stevens is, as of 2008, the oldest and longest-serving member of the Supreme Court, and is generally considered to be the leader of its liberal faction. In the following opinion, Stevens agrees with the Court's decision regarding the case decided concurrently with United States v. Usery, *since that case involved forfeiture of drug money, but disagrees strongly about the seizure of Ursery's home. Contrary to what the Court majority maintained, he says, the decision that it was not punishment was not consistent with the precedents set by earlier cases. He argues that these cases established no general rule that forfeiture is not punitive, and that the opinions in some of them specifically implied that it sometimes is. The majority in this case held that the forfeiture was remedial—that is, that it served to remedy harm done by unlawful activity—but Stevens cannot see how a home in which marijuana was found has substantially facilitiated a narcotics offense, or why forfeiture of it would result in reduction of the drug trade. In his opinion, the seizure was obviously intended as punishment and, therefore, because Usery was sen-*

John Paul Stevens, opinion concurring in part and dissenting in part, *United States v. Ursery*, U.S. Supreme Court, June 24, 1996.

tenced to prison, the double jeopardy clause applies. He suggests that if during the Prohibition era, the government had seized every house in which alcohol was consumed on the pretense that it was remedial, that would have been recognized as far too drastic.

The question the Court poses is whether civil forfeitures constitute "punishment" for purposes of the Double Jeopardy Clause. Because the numerous federal statutes authorizing forfeitures cover such a wide variety of situations, it is quite wrong to assume that there is only one answer to that question. For purposes of analysis, it is useful to identify three different categories of property that are subject to seizure: proceeds, contraband, and property that have played a part in the commission of a crime. The facts of these two cases illustrate the point.

In [the Arlt and Wren case] the Government has forfeited $405,089.23 in currency. Those funds are the proceeds of unlawful activity. They are not property that respondents have any right to retain. The forfeiture of such proceeds, like the confiscation of money stolen from a bank, does not punish respondents, because it exacts no price in liberty or lawfully derived property from them. I agree that the forfeiture of such proceeds is not punitive, and therefore I concur in the Court's disposition of [that case].

None of the property seized in [the Ursery case] constituted proceeds of illegal activity. Indeed, the facts of that case reveal a dramatically different situation. Respondent Ursery cultivated marijuana in a heavily wooded area not far from his home in Shiawassee County, Michigan. The illegal substance was consumed by members of his family, but there is no evidence, and no contention by the Government, that he sold any of it to third parties. Acting on the basis of the incorrect assumption that the marijuana plants were on respondent's property, Michigan police officers executed a warrant to search the premises. In his house, they found marijuana seeds, stems, stalks, and a grow light. I presume those items were seized, and I have no difficulty concluding that such a seizure does not constitute punishment, because respondent had no right to possess contraband. Accordingly, I agree with the Court's opinion insofar as it explains why the forfeiture of contraband does not constitute punishment for double jeopardy purposes.

The critical question presented in [the Ursery case] arose not out of the seizure of contraband by the Michigan police, but rather out of the decision by the United States Attorney to take respondent's home. There is no evidence that the house had been purchased with the proceeds of unlawful activity, and the house itself was surely not contraband. Nonetheless, [the law] authorized the Government to seek forfeiture of respondent's residence because it had been used to facilitate the manufacture and distribution of marijuana. Respondent was then himself prosecuted for and convicted of manufacturing marijuana. In my opinion, none of the reasons supporting the forfeiture of proceeds or contraband provides a sufficient basis for concluding that the confiscation of respondent's home was not punitive. . . .

The Court's Decision Disregards Precedents

In recent years, both Congress and the state legislatures have armed their law enforcement authorities with new powers to forfeit property that vastly exceed their traditional tools. In response, this Court has reaffirmed the fundamental proposition that all forfeitures must be accomplished within the constraints set by the Constitution. This Term, the Court has begun dismantling the protections it so recently erected. In *Bennis v. Michigan* (1996), the Court held that officials may confiscate an innocent person's automobile. And today, for the first time, it upholds the forfeiture of a person's home. On the way to its surprising conclusion that the owner is not punished by the loss of his residence, the Court repeatedly professes its adherence to tradition and time-honored practice. As

I discuss below, however, the decision shows a stunning disregard not only for modern precedent, but for our older ones as well.

In the Court's view, the seminal case is *Various Items of Personal Property v. United States* (1931), which approved the forfeiture of an illegal distillery by resort to the "legal fiction" that the distillery, rather than its owner, was being punished "as though it were conscious, instead of inanimate and insentient." Starting from that fanciful premise, the Court was able to conclude that confiscating the property after the owner was prosecuted for the underlying violations of the revenue laws did not offend the Double Jeopardy Clause.

According to the Court, *Various Items* established a categorical rule that the Double Jeopardy Clause was "inapplicable to civil forfeiture actions." The Court asserts that this rule has received "remarkably consistent" application and was "reaffirmed" by a pair of cases in 1972 and 1984. In reality, however, shortly after its announcement, *Various Items* simply disappeared from our jurisprudence. We cited that case in only two decisions over the next seven years, and never again in nearly six decades. Neither of the two cases that supposedly "affirmed" *Various Items—One Lot Emerald Cut Stone v. United States* (1972), and *United States v. One Assortment of 89 Firearms* (1984)—even mentioned it.

More important, neither of those cases endorsed the asserted categorical rule that civil forfeitures never give rise to double jeopardy rights. Instead, each carefully considered the nature of the particular forfeiture at issue, classifying it as either "punitive" or "remedial," before deciding whether it implicated double jeopardy. *Emerald Cut Stones* concerned a Customs statute that authorized confiscation of certain merchandise, in that case jewelry, that had been smuggled into the United States. The Court explained that the purpose of the statute was to remove such items from circulation, and that the penalty amounted to a reasonable liquidated damages award to reimburse the Government for the costs of enforcement and investigation. In those respects, therefore, it constituted a "remedial rather than punitive sanctio[n]." In *89 Firearms*, the Court explored in even greater detail the character of a federal statute that forfeited unregistered firearms. It reasoned that the sanction "further[ed] broad remedial aims" in preventing commerce in such weapons, and also covered a broader range of conduct than simply criminal behavior. For those reasons, it was not properly characterized as a punitive sanction. . . .

In *Boyd v. United States* (1886), the Court held that compulsory production of an individual's private papers for use in a proceeding to forfeit his property for alleged fraud against the revenue laws violated both the Fourth Amendment and the Fifth Amendment's Self-Incrimination Clause. As the Court stated, "proceedings instituted for the purpose of declaring the forfeiture of a man's property by reason of offences committed by him, though they may be civil in form, are in their nature criminal," and thus give rise to these constitutional safeguards.

We reaffirmed *Boyd* twice during the span of time between our decisions in *Various Items* and *89 Firearms*. In *One 1958 Plymouth Sedan v. Pennsylvania* (1965), the Court unanimously repeated *Boyd's* conclusion that "a forfeiture proceeding is quasi-criminal in character" and "[i]ts object, like a criminal proceeding, is to penalize for the commission of an offense against the law." The Court therefore held that the Fourth Amendment applied to a proceeding to forfeit an automobile used to transport illegally manufactured liquor.

Even more significant is *United States v. United States Coin & Currency* (1971), in which the Court again held that the Fifth Amendment applied to forfeiture proceedings. . . .

Emerald Cut Stones expressly recognized the continuing validity of *Coin & Currency* and *One 1958 Plymouth Sedan*. It distinguished the Customs statute in that case because the forfeiture did not depend on the fact of a criminal offense or conviction. That recognition is critical. For whatever its connection to the Excessive Fines Clause of the Eighth Amendment, the Double Jeopardy Clause is part of the same Amendment as the Self-Incrimination Clause, and ought to be interpreted *in pari materia* [in the light of that clause, because it deals with the same subject]. By confining its holding to civil forfeitures fairly characterized as remedial, and by distinguishing cases that had applied the Fifth Amendment to other types of forfeitures, *Emerald Cut Stones* and *89 Firearms* recognized the possibility that the Double Jeopardy Clause might apply to certain punitive civil forfeiture proceedings. One of the mysteries of the Court's opinion is that, although it claims that civil *in rem* forfeiture cannot be understood as punishment, it devotes Part II-C to examining the actual purposes of the forfeiture in this case and "proving" that they are not punitive. If the Court truly adhered to the logic of its position, that entire section would be unnecessary. . . .

Forfeiture of a Home Is Not Remedial

The Court supposes that forfeiture of respondent's house is remedial in nature because it was an instrumentality of a drug crime. It is perfectly conceivable that certain kinds of instruments used in the commission of crimes could be forfeited for remedial purposes. Items whose principal use is illegal—for example, the distillery in *Various Items*—might be thus forfeitable. But it is difficult to understand how a house in which marijuana was found helped to substantially "facilitate" a narcotics offense, or how forfeiture of that house will meaningfully thwart the drug trade. In *Austin* [*v. United States* (1993)] we rejected the argument that a mobile home and body shop were "instruments" of drug trafficking simply because marijuana was sold out of them. I see no basis for a distinction here.

Second, the Court claims that the statute serves the purpose of deterrence, which helps to show that it is remedial, rather than punitive, in character. That statement cannot be squared with our precedents. . . .

For good measure, the Court also rejects two considerations that persuaded the majority in *Austin* to find [the forfeiture law] a punitive statute. The Court first asserts that the statute contains no *scienter* [knowledge of wrongdoing] requirement and property may be forfeited summarily if no one files claim to it. Property that is not claimed, however, is considered abandoned; it proves nothing that the Government is able to forfeit property that no one owns. Any time the Government seeks to forfeit claimed property, it must prove that the claimant is culpable, for the statute contains an express "innocent owner" exception. Today the Court finds the structure of the statute irrelevant, but *Austin* said that the exemption for innocent owners "makes [the statute] look more like punishment." . . .

Finally, the Court announces that the fact that the statute is "tied to criminal activity" is insufficient to render it punitive. *Austin* expressly relied on Congress' decision to "tie forfeiture directly to the commission of drug offenses" as evidence that it was intended to be punitive.

The recurrent theme of the Court's opinion is that there is some mystical difference between *in rem* and *in personam* proceedings, such that only the latter can give rise to double jeopardy concerns. The Court claims that, "[s]ince at least *Various Items*," we have drawn this distinction for purposes of applying relevant constitutional provisions. That statement, however, is incorrect. We have repeatedly rejected the idea that the nature of the court's jurisdiction has any bearing on the constitutional protections that apply at a proceeding before it. . . .

The pedantic distinction between *in rem* and *in personam* actions is ultimately only a cover for the real basis for the Court's decision: the idea that the property, not the owner, is being "punished" for offenses of which it is "guilty." Although the Court prefers not to rely on this notorious fiction too blatantly, its repeated citations to *Various Items* make clear that the Court believes respondent's home was "guilty" of the drug offenses with which he was charged. On that rationale, of course, the case is easy. The *owner* of the property is not being punished when the Government confiscates it, just the *property*. The same sleight-of-hand would have worked in *Austin*, too: The owner of the property is not being excessively fined, just the property itself. Despite the Government's heavy reliance on that fiction in *Austin*, we did not allow it to stand in the way of our holding that the seizure of property may punish the owner.

Even if the point had not been settled by prior decisions, common sense would dictate the result in this case. There is simply no rational basis for characterizing the seizure of this respondent's home as anything other than punishment for his crime. The house was neither proceeds nor contraband, and its value had no relation to the Government's authority to seize it. Under the controlling statute, an essential predicate for the forfeiture was proof that respondent had used the property in connection with the commission of a crime. The forfeiture of this property was unquestionably "a penalty that had absolutely no correlation to any damages sustained by society or to the cost of enforcing the law" [*United States v. Ward* (1980)]. As we unanimously recognized in [United States v.] *Halper* [(1989)], formalistic distinctions that obscure the obvious practical consequences of governmental action disserve the "'humane interests'" protected by the Double Jeopardy Clause. Fidelity to both reason and precedent dictates the conclusion that *this forfeiture* was "punishment" for purposes of the Double Jeopardy Clause. . . .

Congress' decision to create novel and additional penalties should not be permitted to eviscerate the protection against governmental overreaching embodied in the Double Jeopardy

Clause. That protection has far deeper roots than the relatively recent enactments that have so dramatically expanded the sovereign's power to forfeit private property.

One final example may illustrate the depth of my concern that the Court's treatment of our cases has cut deeply into a guarantee deemed fundamental by the Founders. The Court relies heavily on a few early decisions that involved the forfeiture of vessels whose entire mission was unlawful and on the Prohibition era precedent sustaining the forfeiture of a distillery—a property that served no purpose other than the manufacture of illegal spirits. Notably none of those early cases involved the forfeiture of a home as a form of punishment for misconduct that occurred therein. Consider how drastic the remedy would have been if Congress in 1931 had authorized the forfeiture of every home in which alcoholic beverages were consumed. Under the Court's reasoning, I fear that the label "civil," or perhaps "*in rem*," would have been sufficient to avoid characterizing such forfeitures as "punitive" for purposes of the Double Jeopardy Clause. Our recent decisions in *Halper, Austin*, and [*Department of Revenue of Mont v.*] *Kurth Ranch*, dictate a far different conclusion. I remain persuaded that those cases were correctly decided and should be followed today.

> *"In its* Ursery *decision, the Supreme Court not only turned a blind eye to injustice, but also to the corrupt influence of asset seizures on police and prosecutors."*

Civil Forfeiture Laws Are Not Fair and Have Not Reduced Drug Use

Paul Craig Roberts

Paul Craig Roberts is a nationally syndicated columnist for The Washington Times. *In the following viewpoint, he argues that because the police and prosecutors have a financial stake in seized property, forfeiture laws have corrupted some of them. They look for cash and simply assume that it is drug money. The government does not need evidence to seize money or property under these laws, and it does not need to establish the guilt of people from whom it is taken. Forfeiture laws have not reduced drug use, Roberts says, but they have made thieves out of law enforcement officers.*

The Supreme Court seems to think that anything is permissible in the name of fighting crime. On June 24, 1996, *United States vs. Ursery*, the court ruled that civil punishments in addition to criminal punishments do not constitute double jeopardy. The court could have avoided such a dangerous ruling by studying Pulitzer Prize-winner Leonard W. Levy's new book, *A License to Steal: The Forfeiture of Property*.

Paul Craig Roberts, "Legalized Larceny Posing as Forfeiture," *Washington Times*, July 10, 1996, p. 17. Copyright © 1996 News World Communications, Inc. Reproduced by permission.

Asset forfeitures came to prominence in the war against drugs. They have not dented drug use, but they have made thieves out of law enforcement officers. Mr. Levy recounts how Suffolk County, N.Y., district attorney James M. Catterson, drives a "swanky BMW as his official car instead of a county car." The luxury import was part of $3 million of property seized by Mr. Catterson.

Somerset County, N.J., prosecutor Nicholas L. Bissell used $6,000 of seized funds "for a corporate membership in a private tennis and health club for the benefit of his 17 assistant prosecutors and 50 detectives."

Donald A. Regan of Montvale, N.J., made the mistake of giving a lift to an individual who had cocaine in his possession. Mr. Regan and his passenger were arrested, but the charges against him were dropped when the passenger swore to his innocence. But the police decided to keep Mr. Regan's car as an "instrumentality" of crime.

Forfeiture powers have made police officers more corrupt than the Sheriff of Nottingham [the villain in the story of Robin Hood]. Mr. Levy reports that police focus on cars with cash, not on cars with caches of drugs: "The purpose is to hunt for cash." In Volusia County, Fla., police have confiscated $8 million in cash from southbound drivers on Interstate 95, because the police presume that cash indicates a desire to purchase drugs.

After Hurricane Hugo, Selena Washington and Willie Jones drove south on I-95 to get construction materials to repair her damaged home. Being black, they fit a "drug carrier profile" and were stopped. The police officer took $19,000 from Mrs. Washington's purse, alleging that it was drug money. Without even taking her name, he drove off with his loot.

Mrs. Washington hired an attorney who advised her to share her money with the police because of the legal futility of her position. A deal was struck, and the sheriff kept $4,000.

In order to seize property, the government not only doesn't need to establish guilt, but also doesn't need any facts. As Mr. Levy states, "Hearsay, circumstantial evidence, and anything more than a hunch can be used to establish probable cause. That done, the burden of proof shifts to the owner of the property or its claimant who must establish by a preponderance of evidence that the property was unconnected to a crime. In effect, he must prove the innocence of his property." This is often difficult, because 90 percent of the money in circulation tests positive for cocaine. Any police dog can sniff almost anyone's wallet and wag its tail to establish probable cause.

Because law enforcement officers have been given a financial stake in seized property, raising cash becomes more important than catching criminals. Mr. Levy quotes one Justice Department official: "Increasingly, you're seeing supervisors of cases saying, 'Well, what can we seize?' when they're trying to decide what to investigate. . . . They're paying more attention to the revenues they can get . . . and it's skewing the cases they get involved in."

In its *Ursery* decision, the Supreme Court not only turned a blind eye to injustice, but also to the corrupt influence of asset seizures on police and prosecutors. Under this ruling, a person can prevail against a government prosecutor in a criminal trial, only to have the prosecutor seize the person's property in a "civil forfeiture."

Justice John Paul Stevens dissented, noting that asset forfeitures increasingly lack any semblance of proportionality to the offense. Guy Jerome Ursery had his house seized because police found a few marijuana seeds, stems and stalks—no growing plants—on his property. No evidence existed that Mr. Ursery's house was bought with the proceeds of drug money.

To bolster its claim that the seizure of Mr. Ursery's house did not constitute punishment, the court's majority relied on ancient cases involving the seizure of pirate ships. "Notably,"

Justice Stevens said, "none of those early cases involved the forfeiture of a home as a form of punishment for misconduct that occurred therein. Consider how drastic the remedy would have been if Congress in 1931 had authorized the forfeiture of every home in which alcoholic beverages were consumed." The [conservative William] Rehnquist court would have blessed such a constitutional violation.

> "Law enforcement is constitutionally permitted to use civil forfeiture, undeterred by double jeopardy concerns, to seek out and confiscate the ill-gotten gains of criminals and to dismantle their criminal organizations."

Civil Forfeiture Laws Can Take the Profit out of Crime

William R. Schroeder

At the time this article was written, William R. Schroeder was a special agent of the FBI and chief of the Legal Forfeiture Unit, Office of General Counsel, at FBI Headquarters. In the following viewpoint, he explains that since the Supreme Court's decision in United States v. Ursery, *federal law enforcement agencies have been encouraged to make greater use of civil forfeiture proceedings, which, he says, they were previously reluctant to do. It was feared that the courts might view civil forfeiture as punishment and that therefore, because the Constitution prohibits double jeopardy (being punished twice for the same crime) offenders whose property had been seized could not be sent to prison. The* Ursery *decision eliminated this possibility by specifically ruling that civil forfeiture could not result in double jeopardy. Also, Schroeder says, it was thought that civil forfeiture might be considered unfair because it allowed property to be seized from people who did not know how their property was being used. However, in a case preceding* Ursery, *the Supreme Court held*

William R. Schroeder, "Civil Forfeiture," *FBI Law Enforcement Bulletin*, vol. 65, October 1996, pp. 28–32.

that innocence of a property owner has not been considered relevant in the past and that allowing seizure of such property makes people more careful about letting it be used by others.

On February 12, 1996, the Attorney General of the United States issued a directive that urged all federal prosecutors and agents of the Federal Bureau of Investigation (FBI), Drug Enforcement Administration (DEA), and Immigration and Naturalization Service (INS) to reinvigorate their efforts in using asset forfeiture as a law enforcement tool. The Attorney General sought this additional commitment because asset forfeiture has proven to be one of the most effective methods available in the continuing battle against major drug traffickers, organized crime figures, and their criminal organizations. As the Attorney General pointed out in the directive, the failure to attack the economic infrastructure of criminals and their organizations seriously limits the effect that any prosecution of these criminals would have on improving the safety and welfare of the American public.

A decline in the use of asset forfeiture by federal law enforcement over the past two years [1994–1996] prompted this reinvigoration effort. Although a number of factors contributed to this decline, the loss of asset forfeiture's effectiveness in addressing serious crime and criminals is not one of them.

The forfeiture of property stifles the goals of those who, motivated by greed, engage in criminal activity. Forfeiture takes the profit out of crime, deprives the criminal of the money essential to finance future criminal conduct, and works to dismantle the financial underpinnings of the criminal organization.

This article discusses two recent Supreme Court cases affecting the use of civil forfeiture. The first case addresses one of the major reasons for the downward spiral in the use of assest forfeiture—the concern over double jeopardy. The second case addresses another issue that has contributed to the decline—the perception that civil forfeiture is unfair.

Double Jeopardy

One major reason for the decline in asset forfeiture has been the concern that its use may bar subsequent criminal prosecution. Civil forfeiture, unlike its criminal counterpart, involves procedures that allow pretrial seizure of assets, requires a relatively low burden of proof on the part of the government, and is not contingent on an owner's conviction. These factors lead to the basis for the concern—whether the combination of criminal prosecution and civil forfeiture may be successive punishment for the same crime, in violation of the Double Jeopardy Clause of the Fifth Amendment.

In the past two years, a significant number of opinions issued by the courts focused on this dilemma. Two lower court rulings, in particular, led to the Supreme Court decision regarding double jeopardy discussed in this article. An adverse decision by the Court could have freed hundreds of drug dealers across the United States, required the return of millions of dollars of ill-gotten gain, and prevented the future use of one of the most important law enforcement tools available to attack the organizational infrastructure of criminal cartels.

In September 1994, the U.S. Court of Appeals for the Ninth Circuit reversed a forfeiture order against various properties valued at approximately $1 million that had been seized from convicted methamphetamine dealers. The court found that the civil forfeiture of drug proceeds and property used in money laundering, which followed the dealers' convictions on drug trafficking and money laundering charges, constituted "punishment" under the Double Jeopardy Clause of the Fifth Amendment.

The court based its finding in large measure on two recent Supreme Court decisions, one involving double jeopardy and the other involving the Excessive Fines Clause of the Eighth Amendment. Both of these cases involved findings by the Court that certain government civil sanctions, despite the "civil" label, could constitute punishment under the Constitution and, therefore, require greater protections. The Ninth Circuit panel reasoned that because the civil forfeiture action was the second punishment imposed for the same offense, the Double Jeopardy Clause required them to reverse the forfeiture order and return the property.

In the other case, decided by the U.S. Court of Appeals for the Sixth Circuit, the government and Guy Ursery had agreed to settle a forfeiture action filed against his residence after the Michigan State Police discovered marijuana and marijuana-growing paraphernalia at the residence. After the property, which had been used to facilitate the unlawful distribution of a controlled substance, was forfeited under federal drug laws, Ursery was convicted of manufacturing marijuana and sentenced to five years' imprisonment. The appellate court reversed Ursery's conviction, concluding that the Double Jeopardy Clause of the Fifth Amendment protects against multiple punishments and prohibits the government from punishing twice for the same offense.

Supreme Court Finds Civil Forfeiture Is Remedial, Not Punitive

In *United States v. Ursery*, a decision filed on June 24, 1996, the U.S. Supreme Court resolved the concern over double jeopardy in favor of law enforcement. The Court consolidated the two lower court cases for purposes of its opinion and, in an 8-to-1 ruling, demonstrated a strong consensus regarding the role of civil forfeiture in law enforcement.

In reversing both lower court decisions, the Supreme Court ruled that the government could use, in combination, the criminal law to prosecute someone and the civil forfeiture laws to confiscate that person's property, even where both actions were based upon the same underlying criminal offense. The Court held these "in rem [action against property rather than its owner] civil forfeitures are neither 'punishment' nor criminal for purposes of the Double Jeopardy Clause." In its analysis, the Court reviewed "a long line of cases [in which the] Court has considered the application of the Double Jeopardy Clause to civil forfeitures, consistently concluding that the Clause does not apply to such actions because they do not impose punishment." The Court concluded that for over fifty years, *in rem* forfeiture has been found to be a remedial sanction, distinct and different from other potentially punitive *in personam* civil penalties, and does not constitute a punishment under the Double Jeopardy Clause.

In 1984, the Supreme Court established a two-stage analysis to determine whether a forfeiture was punishment under the Double Jeopardy Clause or was remedial in nature and not subject to the multiple punishment prong of the clause. This analysis focused on whether Congress intended to establish a civil remedial sanction and whether, despite such intent, the civil forfeiture was so punitive in purpose or effect that its application constituted punishment. Because nothing in its decisions had replaced this traditional understanding, the Court used this same two-staged analysis and concluded that the civil forfeiture actions in question were remedial.

In the first stage of its analysis, the Court found "there was little doubt" that the laws involved in the forfeiture actions in question were intended by Congress to be civil actions, distinct from criminal sanctions *in personam*. The Court supported this finding with the fact that Congress incorporated the *in rem* procedures of the customs laws, which target the property rather than the owner and include procedures for administrative forfeiture actions.

The finding that Congress intended the civil forfeitures in these cases to be *in rem* actions established a presumption that they do not implicate double jeopardy. The Court noted that when Congress has intended a forfeiture action to be civil and, therefore, remedial, only the "clearest proof" that it is "so punitive either in purpose or effect" will overcome the presumption that double jeopardy does not apply.

In the second stage, the Court found little evidence that the forfeitures under these statutes were so punitive, in form or effect, to render them punishment. The Court stated that the most significant factor in reaching this conclusion was that such civil forfeitures serve important nonpunitive goals. The important goals of civil forfeiture include encouraging property owners to take greater care in managing their property, ensuring that owners will not permit their property to be used for illegal purposes, and ensuring that "persons do not profit from their illegal acts."

Forfeiture of Jointly Owned Property Upheld

Another reason for the decline in the use of asset forfeiture is the troubling perception that civil forfeiture can be unfair. This perception is based, in part, on the fact that a civil forfeiture of property can be obtained against an owner who was not involved in or aware of the criminal activity in which the property was involved.

In *Bennis v. Michigan* [1996], the Supreme Court examined such a forfeiture action under the light of constitutional scrutiny and upheld the confiscation. In *Bennis*, the Court found that a state forfeiture law that permitted forfeiture of property from an owner who was unaware of its illegal use by another in lawful possession of the property did not offend the constitutional rights of that owner.

Detroit police had arrested John Bennis after observing him engaged in a sexual act with a prostitute in his car, which was jointly owned by Bennis and his wife. After Bennis was convicted of gross indecency, the state had the car declared a public nuisance and abated (forfeited) under state law.

The Circuit Court of Wayne County rejected the wife's challenge to the forfeiture on the grounds that she was innocent of any wrongdoing and that she had no knowledge her husband would use the vehicle to violate the law. Subsequently, on appeal, the Michigan Supreme Court upheld the forfeiture, finding that the state law did not provide for an innocent owner defense, and therefore, the state did not need to prove the owner knew or agreed that the car would be used illegally.

The U.S. Supreme Court reviewed the case in order to determine whether Michigan's forfeiture of the vehicle deprived the wife of her interest in the car without due process in violation of the Fourteenth Amendment, or if it amounted to a taking of her interest for public use without compensation in violation of the Fifth Amendment, as incorporated by the Fourteenth Amendment. The Court affirmed the decision of the Michigan Supreme Court and upheld the forfeiture.

"Innocent Owner" Defense

Citing a long and unbroken line of cases holding that an owner's interest in property may be forfeited by reason of the property's use, even though the owner was unaware of that use, the Court concluded that the issue is "too firmly fixed in the punitive and remedial jurisprudence of the country to be now displaced." The Court relied on the rationale of earlier decisions to find that even though Mrs. Bennis did not know her car would be used illegally, its forfeiture by the state did not offend the Due Process Clause of the Fourteenth Amendment. The holdings of the earlier cases cited by the Court have generally been based on the concept that in a civil forfeiture it is the property itself that is being held accountable for its illegal use and thus the "innocence" of the owner is not relevant.

In its decision, the Supreme Court noted that the innocence of an owner has almost uniformly been rejected as a defense to forfeiture. Additionally, the Court has consistently upheld the forfeiture of property from owners "without guilt" because such laws furthered "the punitive and deterrent purposes" of the legislature in enacting laws that "may have the desirable effect of inducing them (owners) to exercise greater care in transferring possession of their property." The Court answered the Fifth Amendment "taking" question raised by Bennis by finding that the government may not be required to compensate an owner for property that has been acquired through a civil forfeiture action.

The Court, in reaching its conclusion, noted that civil forfeiture imposes an economic penalty on illegal use that is neither unconstitutional nor unfair. The Court balanced the increased responsibility for property owners against the need to deter criminal conduct and decided in favor of law enforcement. In this regard, civil forfeiture is no different than other laws based on important public interest precepts.

Civil forfeiture of property serves both remedial and deterrent purposes that are distinct from any punishment that may result from criminal prosecution of the owner. Such forfeitures help prevent future illegal use and impose an economic disincentive, rendering the behavior unprofitable.

The decision by the Court in *Ursery* permits federal and state law enforcement to resume the seizure and forfeiture of property under the civil procedures without fear that such action will bar subsequent prosecution of the owner. It also provides a clear signal that the Supreme Court recognizes and supports the strong "nonpunitive goals" of civil forfeiture that serve law enforcement in combating crime.

The significance of *Bennis*, however, lies more in the rationale employed by the Court than in the outcome. It was a precursor to the Court's analysis in *Ursery*, and returned the historical perspective of *in rem* forfeiture actions that seemed to have been changed in *Austin v. United States* [1993].

The *Bennis* decision should not be seen, however, as an unqualified justification for forfeiture where the owner is known to be innocent of any wrongdoing. In *Bennis*, an associate justice in a concurring opinion warned that "[i]mproperly used, forfeiture could become more like a roulette wheel employed to raise revenue from innocent but hapless owners whose property is unforeseeably misused, or a tool wielded to punish those who associate with criminals, than a component of a system of justice." He noted that the Constitution assigns to states and federal agencies that use the forfeiture tool the primary responsibility for avoiding that unwanted result. Failure to use this authority consistent with its purpose could result in legislative disapproval.

The seizure of property from one who is truly innocent does not serve the goals of civil forfeiture, i.e., taking the profit out of crime and acting as a deterrent to future criminal activity. In fact, Congress has provided an innocent owner exception to most federal civil forfeiture laws that would, under similar facts, have allowed Bennis to recover her interest in the car by showing her lack of knowledge of the intended illegal use of the vehicle by her wayward husband, and her lack of consent to that usage.

As a result of these Supreme Court decisions, law enforcement is constitutionally permitted to use civil forfeiture, undeterred by double jeopardy concerns, to seek out and confiscate the ill-gotten gains of criminals and to dismantle their criminal organizations. Both federal and state law enforcement agencies working together will be able to use civil forfeiture to effectively take the profit out of crime and deter future criminal conduct motivated by greed.

CHAPTER 3

Drug Testing
of Participants
in Extracurricular
Activities Is Legal

Case Overview

Pottawatomie v. Earls (2002)

Lindsay Earls and Daniel James, both students at Tecumseh High School in Oklahoma, were active in extracurricular activities. Lindsey was a member of the show choir, the marching band, the Academic Team, and the National Honor Society. Daniel wanted to join the Academic Team. In 1998, the Board of Education of Pottawatomie County, which administers the Tecumseh schools, adopted a policy requiring all middle and high school students to consent to random drug testing in order to participate in any extracurricular activity. Neither Lindsay nor Daniel used illegal drugs and they objected to having to submit to urine tests—which Lindsay called "humiliating and accusatory"—merely because they were members of groups such as those focused on academics and music. Together with their parents they sued the school district, arguing that the policy was a violation of the Constitution's Fourth Amendment prohibition of unreasonable searches.

A previous Supreme Court ruling, *Vernonia v. Acton* (1995), had made drug testing of school athletes legal, so on that basis, the lower court ruled in favor of the school district. It noted that there are special needs in public schools and that, although Tecumseh High did not have a serious drug problem, there was a history of drug abuse there that presented a legitimate cause for concern. The Court of Appeals, however, held that because the school district had failed to demonstrate a drug problem among students participating in extracurricular activities, the policy was indeed unconstitutional. The school district then appealed that decision to the U.S. Supreme Court.

Major health and education authorities that oppose drug testing became involved in the case. A joint amici curiae (friend of the court) brief submitted by the American Academy of Pediatrics, National Education Association, American Public Health Association, and others, stated, "There is growing recognition that extracurricular involvement plays a role in *protecting* students from substance abuse and other dangerous health behaviors. A policy that conditions participation on submission to an in-school urine testing regime not only risks denying those benefits to students who, out of principle or modesty, refuse to submit, but it is disproportionately likely to discourage 'marginal,' higher risk students."

Nevertheless, in a five to four decision the Supreme Court ruled that Tecumseh's drug testing program did not violate the Fourth Amendment. In addition to the reasons stated in the Court's majority opinion, Justice Stephen Breyer wrote in a concurring opinion that drug testing "offers the adolescent a nonthreatening reason to decline his friends' drug-use invitations, namely that he intends to play baseball, participate in debate, join the band or engage in any one of a half-dozen useful, interesting and important activities."

Many people who had been wanting schools to do more about the drug problem were pleased with the ruling. But others agreed with the National Association of Social Workers, which stated, "The Supreme Court's decision today to allow drug testing in schools sets a precedent, no matter how well-intentioned, that social workers believe is both invasive and counterproductive to combating drug and alcohol abuse in schools."

Although the ruling in *Pottawatomie v. Earls* made random drug testing in schools legal under federal law, this does not mean that all schools are free to do it. Schools must abide by state law as well as federal law, and some states have constitutions that offer more privacy protection than the U.S. Constitution. Therefore, a number of policies similar to Tecumseh's have been struck down. Nevertheless, the George W. Bush administration has allocated millions of dollars to grants for school drug testing and has continued to urge that more schools adopt such programs.

"The nationwide drug epidemic makes the war against drugs a pressing concern in every school."

The Court's Decision: Suspicionless Drug Testing Is a Reasonable Means of Deterring Drug Use Among Schoolchildren

Clarence Thomas

Clarence Thomas, the second African American to serve on the Supreme Court, has been a justice since 1991. He is among the Court's most conservative members. In the following, presenting the majority decision in Pottawatomie v. Earls, *Thomas argues that drug testing of all students who participate in extracurricular activities is reasonable and thus is not barred by the Fourth Amendment. Schoolchildren have less expectation of privacy than adults, he says, and, in any case, the level of privacy invasion required by the testing is minimal. The results of the tests are not turned over to law enforcement agencies; the worst that can happen to a student who tests positive is to be barred from extracurricular activities. Thomas asserts that those who choose to participate in after-school activities are subject to many other requirements that do not apply to the whole student body. Furthermore, the Fourth Amendment does not require that there be any individualized suspicion of wrongdoing before a search can*

Clarence Thomas, majority opinion, *Board of Education of Independent School District No. 92 of Pottawatomie County v. Earls*, U.S. Supreme Court, June 27, 2002.

be conducted. The Court held that there is just as much reason to test students involved in other activities as to test athletes, which had been ruled constitutional in an earlier case.

The Fourth Amendment to the United States Constitution protects "[t]he right of the people to be secure in their persons, houses, papers, and effects, against unreasonable searches and seizures." Searches by public school officials, such as the collection of urine samples, implicate Fourth Amendment interests. We must therefore review the [Pottawatomie County] School District's Policy for "reasonableness," which is the touchstone of the constitutionality of a governmental search.

In the criminal context, reasonableness usually requires a showing of probable cause. The probable cause standard, however, "is peculiarly related to criminal investigations" and may be unsuited to determining the reasonableness of administrative searches where the "Government seeks to prevent the development of hazardous conditions" [*Treasury Employees v. Von Raab* (1989)]. The Court has also held that a warrant and finding of probable cause are unnecessary in the public school context because such requirements "'would unduly interfere with the maintenance of the swift and informal disciplinary procedures [that are] needed'" [*Vernonia v. Acton* (1995)].

Given that the School District's Policy is not in any way related to the conduct of criminal investigations, respondents [Lindsay Earls and Daniel James] do not contend that the School District requires probable cause before testing students for drug use. Respondents instead argue that drug testing must be based at least on some level of individualized suspicion. It is true that we generally determine the reasonableness of a search by balancing the nature of the intrusion on the individual's privacy against the promotion of legitimate governmental interests. But we have long held that "the Fourth Amendment imposes no irreducible requirement of [individualized] suspicion" [*United States v. Martinez-Fuerte* (1976)].

"[I]n certain limited circumstances, the Government's need to discover such latent or hidden conditions, or to prevent their development, is sufficiently compelling to justify the intrusion on privacy entailed by conducting such searches without any measure of individualized suspicion" [*Von Raab*]. Therefore, in the context of safety and administrative regulations, a search unsupported by probable cause may be reasonable "when 'special needs, beyond the normal need for law enforcement, make the warrant and probable-cause requirement impracticable'" [*Griffin v. Wisconsin* (1987)].

Significantly, this Court has previously held that "special needs" inhere in the public school context. While schoolchildren do not shed their constitutional rights when they enter the schoolhouse, "Fourth Amendment rights . . . are different in public schools than elsewhere; the "reasonableness" inquiry cannot disregard the schools' custodial and tutelary responsibility for children" [*Vernonia*]. In particular, a finding of individualized suspicion may not be necessary when a school conducts drug testing.

In *Vernonia*, this Court held that the suspicionless drug testing of athletes was constitutional. The Court, however, did not simply authorize all school drug testing, but rather conducted a fact-specific balancing of the intrusion on the children's Fourth Amendment rights against the promotion of legitimate governmental interests. Applying the principles of *Vernonia* to the somewhat different facts of this case, we conclude that Tecumseh's Policy is also constitutional.

We first consider the nature of the privacy interest allegedly compromised by the drug testing. . . .

A student's privacy interest is limited in a public school environment where the State is responsible for maintaining discipline, health, and safety. Schoolchildren are routinely required to submit to physical examinations and vaccinations against disease. Securing order in the school environment sometimes requires that students be subjected to greater controls than those appropriate for adults.

Respondents argue that because children participating in nonathletic extracurricular activities are not subject to regular physicals and communal undress, they have a stronger expectation of privacy than the athletes tested in *Vernonia*. This distinction, however, was not essential to our decision in *Vernonia*, which depended primarily upon the school's custodial responsibility and authority.

In any event, students who participate in competitive extracurricular activities voluntarily subject themselves to many of the same intrusions on their privacy as do athletes. Some of these clubs and activities require occasional off-campus travel and communal undress. All of them have their own rules and requirements for participating students that do not apply to the student body as a whole. For example, each of the competitive extracurricular activities governed by the Policy must abide by the rules of the Oklahoma Secondary Schools Activities Association, and a faculty sponsor monitors the students for compliance with the various rules dictated by the clubs and activities. This regulation of extracurricular activities further diminishes the expectation of privacy among schoolchildren. We therefore conclude that the students affected by this Policy have a limited expectation of privacy.

The Invasion of Students' Privacy Is Not Significant

Next, we consider the character of the intrusion imposed by the Policy. Urination is "an excretory function traditionally shielded by great privacy" [*Skinner v. Railway Labor Executives' Association* (1989)]. But the "degree of intrusion" on one's privacy caused by collecting a urine sample "depends upon the manner in which production of the urine sample is monitored" [*Vernonia*].

Under the Policy, a faculty monitor waits outside the closed restroom stall for the student to produce a sample and must "listen for the normal sounds of urination in order to guard against tampered specimens and to insure an accurate chain of custody." The monitor then pours the sample into two bottles that are sealed and placed into a mailing pouch along with a consent form signed by the student. This procedure is virtually identical to that reviewed in *Vernonia*, except that it additionally protects privacy by allowing male students to produce their samples behind a closed stall. Given that we considered the method of collection in *Vernonia* a "negligible" intrusion, the method here is even less problematic.

In addition, the Policy clearly requires that the test results be kept in confidential files separate from a student's other educational records and released to school personnel only on a "need to know" basis. Respondents nonetheless contend that the intrusion on students' privacy is significant because the Policy fails to protect effectively against the disclosure of confidential information and, specifically, that the school "has been careless in protecting that information: for example, the Choir teacher looked at students' prescription drug lists and left them where other students could see them." But the choir teacher is someone with a "need to know," because during off-campus trips she needs to know what medications are taken by her students. Even before the Policy was enacted, the choir teacher had access to this information. In any event, there is no allegation that any other student did see such information. This one example of alleged carelessness hardly increases the character of the intrusion.

Moreover, the test results are not turned over to any law enforcement authority. Nor do the test results here lead to the imposition of discipline or have any academic consequences. Rather, the only consequence of a failed drug test is to limit the student's privilege of participating in extracurricular activities. Indeed, a student may test positive for drugs twice and still be allowed to participate in extracurricular activities. After the first positive test, the school contacts the student's parent or guardian for a meeting. The student may continue to participate in the activity if within five days of the meeting the student shows proof of receiving drug counseling and submits to a second drug test in two weeks. For the second positive test, the student is suspended from participation in all extracurricular activities for fourteen days, must complete four hours of substance abuse counseling, and must submit to monthly drug tests. Only after a third positive test will the student be suspended from participating in any extracurricular activity for the remainder of the school year, or eighty-eight school days, whichever is longer.

Given the minimally intrusive nature of the sample collection and the limited uses to which the test results are put, we conclude that the invasion of students' privacy is not significant.

Drugs Are a Problem in Tecumseh Schools

Finally, this Court must consider the nature and immediacy of the government's concerns and the efficacy of the Policy in meeting them. This Court has already articulated in detail the importance of the governmental concern in preventing drug use by schoolchildren. The drug abuse problem among our Nation's youth has hardly abated since *Vernonia* was decided in 1995. In fact, evidence suggests that it has only grown worse. As in *Vernonia*, "the necessity for the State to act is magnified by the fact that this evil is being visited not just upon individuals at large, but upon children for whom it has undertaken a special responsibility of care and direction." The health and safety risks identified in *Vernonia* apply with equal force to Tecumseh's children. Indeed, the nationwide drug epidemic makes the war against drugs a pressing concern in every school.

Additionally, the School District in this case has presented specific evidence of drug use at Tecumseh schools. Teachers testified that they had seen students who appeared to be under the influence of drugs and that they had heard students speaking openly about using drugs. A drug dog found marijuana cigarettes near the school parking lot. Police officers once found drugs or drug paraphernalia in a car driven by a Future Farmers of America member. And the school board president reposed that people in the community were calling the board to discuss the "drug situation." We decline to second-guess the finding of the District Court that "[v]iewing the evidence as a whole, it cannot be reasonably disputed that the [School District] was faced with a 'drug problem' when it adopted the Policy."

Respondents consider the proffered evidence insufficient and argue that there is no "real and immediate interest" to justify a policy of drug testing nonathletes. We have recognized, however, that "[a] demonstrated problem of drug abuse . . . [is] not in all cases necessary to the validity of a testing regime," but that some showing does "shore up an assertion of special need for a suspicionless general search program" [*Chandler v. Miller* (1997)]. The School District has provided sufficient evidence to shore up the need for its drug testing program.

Furthermore, this Court has not required a particularized or pervasive drug problem before allowing the government to conduct suspicionless drug testing. For instance, in *Von Raab*, the Court upheld the drug testing of Customs officials on a purely preventive basis, without any documented history of drug use by such officials. In response to the lack of evidence relating to drug use, the Court noted generally that "drug abuse is one of the most serious problems confronting our society today," and that programs to prevent and detect drug use among Customs officials could not be deemed unreasonable. Likewise, the need to prevent and deter the substantial harm of childhood drug use provides the necessary immediacy for a school testing policy. Indeed, it would make little sense to require a school district to wait for a substantial portion of its students to begin using drugs before it was allowed to institute a drug testing program designed to deter drug use.

Given the nationwide epidemic of drug use and the evidence of increased drug use in Tecumseh schools, it was entirely reasonable for the School District to enact this particular drug testing policy. We reject the Court of Appeals' novel test that "any district seeking to impose a random suspicionless drug testing policy as a condition to participation in a school activity must demonstrate that there is some identifiable drug abuse problem among a sufficient number of those subject to the testing, such that testing that group of students will actually redress its drug problem."

Among other problems, it would be difficult to administer such a test. As we cannot articulate a threshold level of drug use that would suffice to justify a drug testing program for schoolchildren, we refuse to fashion what would in effect be a constitutional quantum of drug use necessary to show a "drug problem."

Individual Suspicion of Wrongdoing Is Not Required for Drug Testing

Respondents also argue that the testing of nonathletes does not implicate any safety concerns, and that safety is a "crucial factor" in applying the special needs framework. They contend that there must be "surpassing safety interests" [*Skinner*], or "extraordinary safety and national security hazards" [*Von Raab*], in order to override the usual protections of the Fourth Amendment. Respondents are correct that safety factors into the special needs analysis, but the safety interest furthered by drug testing is undoubtedly substantial for all children, athletes, and nonathletes alike. We know all too well that drug use carries a variety of health risks for children, including death from overdose.

We also reject respondents' argument that drug testing must presumptively be based upon an individualized reasonable suspicion of wrongdoing because such a testing regime would be less intrusive. In this context, the Fourth Amendment does not require a finding of individualized suspicion, and we decline to impose such a requirement on schools attempting to prevent and detect drug use by students. Moreover, we question whether testing based on individualized suspicion in fact would be less intrusive. Such a regime would place an additional burden on public school teachers who are already tasked with the difficult job of maintaining order and discipline. A program of individualized suspicion might unfairly target members of unpopular groups. The fear of lawsuits resulting from such targeted searches may chill enforcement of the program, rendering it ineffective in combating drug use. In any case, this Court has repeatedly stated that reasonableness under the Fourth Amendment does not require employing the least intrusive means, because "[t]he logic of such elaborate 'less restrictive alternative' arguments could raise insuperable barriers to the exercise of virtually all search and seizure powers" [*Martinez-Fuerte*].

Finally, we find that testing students who participate in extracurricular activities is a reasonably effective means of addressing the School District's legitimate concerns in preventing, deterring, and detecting drug use. While in *Vernonia* there might have been a closer fit between the testing of athletes and the trial court's finding that the drug problem was "fueled by the 'role model' effect of athletes' drug use," such a finding was not essential to the holding. *Vernonia* did not require the school to test the group of students most likely to use drugs, but rather considered the constitutionality of the program in the context of the public school's custodial responsibilities. Evaluating the Policy in this context, we conclude that the drug testing of Tecumseh students who participate in extracurricular activities effectively serves the School District's interest in protecting the safety and health of its students.

Within the limits of the Fourth Amendment, local school boards must assess the desirability of drug testing schoolchildren. In upholding the constitutionality of the Policy, we express no opinion as to its wisdom. Rather, we hold only that Tecumseh's Policy is a reasonable means of furthering the School District's important interest in preventing and deterring drug use among its schoolchildren.

"The particular testing program upheld today is not reasonable, it is capricious, even perverse."

Dissenting Opinion: Drug Testing the Student Population Least Likely to Take Drugs Is Unconstitutional

Ruth Bader Ginsburg

Ruth Bader Ginsburg has been a justice of the Supreme Court since 1993, and is one of its most liberal members. In her dissenting opinion in Pottawatomie v. Earls, *she strongly disagrees with the Court's decision. That decision was based largely on precedent set in* Vernonia v. Acton *(1995), where the Court ruled that drug testing of school athletes is constitutional. But, she says, there are many differences between the two cases. Although extracurricular activities are voluntary, they are a key component of school life and are essential for students applying to college. Athletes have less expectation of privacy than participants in other activities and the use of drugs is more physically dangerous to them. Furthermore, the schools in Vernonia had a serious drug problem, while those in the Pottawatomie district did not. And because students who are involved in extracurricular activities are less like to take drugs than other students, Pottawatomie's drug testing policy will deter students from participating in the very activities that discourage drug use. In Justice Ginsburg's opinion, that policy is not reasonable, is imper-*

Ruth Bader Ginsburg, dissenting opinion, *Board of Education of Independent School District No. 92 of Pottawatomie County v. Earls,* U.S. Supreme Court, June 27, 2002.

missible under the Fourth Amendment, and will teach young people to discount the importance of constitutional protections.

Seven years ago [1995] in *Vernonia School Dist. 47J v. Acton,* this court determined that a school district's policy of randomly testing the urine of its student athletes for illicit drugs did not violate the Fourth Amendment. In so ruling, the Court emphasized that drug use "increase[d] the risk of sports-related injury" and that Vernonia's athletes were the "leaders" of an aggressive local "drug culture" that had reached "'epidemic proportions.'" Today, the Court relies upon *Vernonia* to permit a school district with a drug problem its superintendent repeatedly described as "not . . . major," to test the urine of an academic team member solely by reason of her participation in a nonathletic, competitive extracurricular activity— participation associated with neither special dangers from, nor particular predilections for, drug use.

"[T]he legality of a search of a student," this Court has instructed, "should depend simply on the reasonableness, under all the circumstances, of the search" [*New Jersey v. T.L.O.* (1985)]. Although "'special needs' inhere in the public school context," those needs are not so expansive or malleable as to render reasonable any program of student drug testing a school district elects to install. The particular testing program upheld today is not reasonable, it is capricious, even perverse: petitioners' policy targets for testing a student population least likely to be at risk from illicit drugs and their damaging effects. I therefore dissent.

A search unsupported by probable cause nevertheless may be consistent with the Fourth Amendment "when special needs, beyond the normal need for law enforcement, make the warrant and probable-cause requirement impracticable" [*Griffin v. Wisconsin* (1987)]. In *Vernonia*, this Court made clear that "such 'special needs' . . . exist in the public school context." . . .

The *Vernonia* Court concluded that a public school district facing a disruptive and explosive drug abuse problem sparked by members of its athletic teams had "special needs" that justified suspicionless testing of district athletes as a condition of their athletic participation.

This case presents circumstances dispositively different from those of *Vernonia*. True, as the Court stresses, Tecumseh students participating in competitive extracurricular activities other than athletics share two relevant characteristics with the athletes of *Vernonia*. First, both groups attend public schools. "[O]ur decision in *Vernonia*," the Court states, "depended primarily upon the school's custodial responsibility and authority." Concern for student health and safety is basic to the school's caretaking, and it is undeniable that "drug use carries a variety of health risks for children, including death from overdose."

Those risks, however, are present for *all* schoolchildren. *Vernonia* cannot be read to endorse invasive and suspicionless drug testing of all students upon any evidence of drug use, solely because drugs jeopardize the life and health of those who use them. Many children, like many adults, engage in dangerous activities on their own time; that the children are enrolled in school scarcely allows government to monitor all such activities. If a student has a reasonable subjective expectation of privacy in the personal items she brings to school, surely she has a similar expectation regarding the chemical composition of her urine. Had the *Vernonia* Court agreed that public school attendance, in and of itself, permitted the State to test each student's blood or urine for drugs, the opinion in *Vernonia* could have saved many words.

Voluntary Nature of Activities Does Not Justify Suspicionless Searches

The second commonality to which the Court points is the voluntary character of both interscholastic athletics and other competitive extracurricular activities. By choosing to "go out for the team," [school athletes] voluntarily subject themselves to a degree of regulation even higher than that imposed on students generally. Comparably, the Court today observes, "students who participate in competitive extracurricular activities voluntarily subject themselves to" additional rules not applicable to other students.

The comparison is enlightening. While extracurricular activities are "voluntary" in the sense that they are not required for graduation, they are part of the school's educational program; for that reason, the petitioner (hereinafter School District) is justified in expending public resources to make them available. Participation in such activities is a key component of school life, essential in reality for students applying to college, and, for all participants, a significant contributor to the breadth and quality of the educational experience.

Voluntary participation in athletics has a distinctly different dimension: Schools regulate student athletes discretely because competitive school sports, by their nature, require communal undress and, more important, expose students to physical risks that schools have a duty to mitigate. For the very reason that schools cannot offer a program of competitive athletics without intimately affecting the privacy of students, *Vernonia* reasonably analogized school athletes to "adults who choose to participate in a closely regulated industry." Industries fall within the closely regulated category when the nature of their activities requires substantial government oversight. Interscholastic athletics similarly require close safety and health regulation; a school's choir, band, and academic team do not.

In short, *Vernonia* applied, it did not repudiate, the principle that "the legality of a search of a student should depend simply on the reasonableness, *under all the circumstances*, of the search." Enrollment in a public school, and election to participate in school activities beyond the bare minimum that the curriculum requires, are indeed factors relevant to reason-ableness, but they do not on their own justify intrusive, suspicionless searches. *Vernonia* accordingly did not rest upon these factors; instead, the Court performed what today's majority aptly describes as a "fact-specific balancing." Balancing of that order, applied to the facts now before the Court, should yield a result other than the one the Court announces today.

Students Who Participate in Nonathletic Activities Expect Privacy

Vernonia initially considered "the nature of the privacy interest upon which the search [there] at issue intrude[d]." The Court emphasized that student athletes' expectations of privacy are necessarily attenuated:

> Legitimate privacy expectations are even less with regard to student athletes. School sports are not for the bashful. They require "suiting up" before each practice or event and show-ering and changing afterwards. Public school locker rooms, the usual sites for these activities, are not notable for the privacy they afford. The locker rooms in *Vernonia* are typi-cal: no individual dressing rooms are provided; shower heads are lined up along a wall, unseparated by any sort of parti-tion or curtain; not even all the toilet stalls have doors. . . . [T]here is an element of communal undress inherent in ath-letic participation.

Competitive extracurricular activities other than athletics, however, serve students of all manner: the modest and shy along with the bold and uninhibited. Activities of the kind plaintiff-respondent Lindsay Earls pursued—choir, show choir, marching band, and academic team—afford opportunities to gain self-assurance, to "come to know faculty members in a less formal setting than the typical classroom," and to acquire "positive social supports and networks [that] play a critical role in periods of heightened stress" [brief for American Acad-emy of Pediatrics].

On "occasional out-of-town trips," students like Lindsay Earls "must sleep together in communal settings and use communal bathrooms." But those situations are hardly equivalent to the routine communal undress associated with athletics; the School District itself admits that when such trips occur, "public-like restroom facilities," which presumably include enclosed stalls, are ordinarily available for changing, and that "more modest students" find other ways to maintain their privacy.

After describing school athletes' reduced expectation of privacy, the *Vernonia* Court turned to "the character of the intrusion . . . complained of." Observing that students produce urine samples in a bathroom stall with a coach or teacher outside, *Vernonia* typed the privacy interests compromised by the process of obtaining samples "negligible." As to the required pretest disclosure of prescription medications taken, the Court assumed that the School District would have permitted [a student] to provide the requested information in a confidential manner—for example, in a sealed envelope delivered to the testing lab. On that assumption, the Court concluded that Vernonia's athletes faced no significant invasion of privacy.

In this case, however, Lindsay Earls and her parents allege that the School District handled personal information collected under the policy carelessly, with little regard for its confidentiality. Information about students' prescription drug use, they assert, was routinely viewed by Lindsay's choir teacher, who left files containing the information unlocked and unsealed, where others, including students, could see them; and test results were given out to all activity sponsors whether or not they had a clear "need to know."

In granting summary judgment to the School District, the District Court observed that the District's "Policy expressly provides for confidentiality of test results, and the Court must assume that the confidentiality provisions will be honored."

The assumption is unwarranted. Unlike *Vernonia*, where the District Court held a bench trial before ruling in the School District's favor, this case was decided by the District Court on summary judgment. At that stage, doubtful matters should not have been resolved in favor of the judgment seeker.

There Was No Serious Drug Problem in Tecumseh Schools

Finally, the "nature and immediacy of the governmental concern," faced by the *Vernonia* School District dwarfed that confronting Tecumseh administrators. *Vernonia* initiated its drug testing policy in response to an alarming situation: "[A] large segment of the student body, particularly those involved in interscholastic athletics, was in a state of rebellion ... fueled by alcohol and drug abuse as well as the student[s'] misperceptions about the drug culture." Tecumseh, by contrast, repeatedly reported to the Federal Government during the period leading up to the adoption of the policy that "types of drugs [other than alcohol and tobacco] including controlled dangerous substances, are present [in the schools], but have not identified themselves as major problems at this time."

As the Tenth Circuit observed, "without a demonstrated drug abuse problem among the group being tested, the efficacy of the District's solution to its perceived problem is ... greatly diminished."

The School District cites *Treasury Employees v. Von Raab*, in which this Court permitted random drug testing of Customs agents absent "any perceived drug problem among Customs employees," given that "drug abuse is one of the most serious problems confronting our society today." The tests in *Von Raab* and [*Skinner v.*] *Railway Labor Executives*, however, were installed to avoid enormous risks to the lives and limbs of others, not dominantly in response to the health risks to users invariably present in any case of drug use.

Not only did the Vernonia and Tecumseh districts confront drug problems of distinctly different magnitudes, they also chose different solutions: Vernonia limited its policy to athletes; Tecumseh indiscriminately subjected to testing all participants in competitive extracurricular activities. Urging that "the safety interest furthered by drug testing is undoubtedly substantial for all children, athletes and nonathletes alike," the Court cuts out an element essential to the *Vernonia* judgment. Citing medical literature on the effects of combining illicit drug use with physical exertion, the *Vernonia* Court emphasized that "the particular drugs screened by [Vernonia's] Policy have been demonstrated to pose substantial physical risks to athletes."

At the margins, of course, no policy of random drug testing is perfectly tailored to the harms it seeks to address. The School District cites the dangers faced by members of the band, who must "perform extremely precise routines with heavy equipment and instruments in close proximity to other students," and by Future Farmers of America, who "are required to individually control and restrain animals as large as 1500 pounds." ...

Notwithstanding nightmarish images of out-of-control flatware, livestock run amok, and colliding tubas disturbing the peace and quiet of Tecumseh, the great majority of students the School District seeks to test in truth are engaged in activities that are not safety sensitive to an unusual degree. There is a difference between imperfect tailoring and no tailoring at all.

The Vernonia district, in sum, had two good reasons for testing athletes: Sports team members faced special health risks and they "were the leaders of the drug culture." No similar reason, and no other tenable justification, explains Tecumseh's decision to target for testing all participants in every competitive extracurricular activity.

Participants in Extracurricular Activities Are Less Likely to Use Drugs

Nationwide, students who participate in extracurricular activities are significantly less likely to develop substance abuse problems than are their less-involved peers. Even if students might be deterred from drug use in order to preserve their extracurricular eligibility, it is at least as likely that other students might forgo their extracurricular involvement in order to avoid detection of their drug use. Tecumseh's policy thus falls short doubly if deterrence is its aim: it invades the privacy of students who need deterrence least and risks steering students at greatest risk for substance abuse away from extracurricular involvement that potentially may palliate drug problems.

To summarize, this case resembles *Vernonia* only in that the School Districts in both cases conditioned engagement in activities outside the obligatory curriculum on random subjection to urinalysis. The defining characteristics of the two programs, however, are entirely dissimilar. The Vernonia district sought to test a subpopulation of students distinguished by their reduced expectation of privacy, their special susceptibility to drug-related injury, and their heavy involvement with drug use. The Tecumseh district seeks to test a much larger population associated with none of these factors. It does so, moreover, without carefully safeguarding student confidentiality and without regard to the program's untoward effects. A program so sweeping is not sheltered by *Vernonia*; its unreasonable reach renders it impermissible under the Fourth Amendment. . . .

It is a sad irony that the petitioning School District seeks to justify its edict here by trumpeting "the schools' custodial and tutelary responsibility for children" [*Vernonia*]. In regulating an athletic program or endeavoring to combat an exploding drug epidemic, a school's custodial obligations may permit searches that would otherwise unacceptably abridge students' rights. When custodial duties are not ascendant, however, schools' tutelary obligations to their students require them to "teach by example" by avoiding symbolic measures that diminish constitutional protections. [As we said in *West Virginia Bd. of Ed. v. Barnette* (1943)] "That [schools] are educating the young for citizenship is reason for scrupulous protection of Constitutional freedoms of the individual if we are not to strangle the free mind at its source and teach youth to discount important principles of our government as mere platitudes."

"Schools across the Nation have a uniform interest in employing drug testing to deter drug use without regard to how much drug use is, in fact, detected."

The Government Has a Compelling Interest in Preventing the Spread of Drug Use in Schools

Theodore B. Olson et al.

Theodore B. Olson was the U.S. Solicitor General from 2001 to 2004. Along with his assistants, he submitted an amicus curiae (friend of the court) brief to the Supreme Court in the case of Pottawatomie v. Earls. *Olson and his colleagues argue that preventing drug use among students who participate in non-athletic extracurricular activities is just as important as preventing it among athletes because there is just as much danger to them, and because they too are role models when they represent their school in the community. Drawing a distinction say these attorneys, is demeaning to both groups of students because it implies that one deserves less protection than the other. A school district should not have to wait for a problem to become serious before taking action, they say. Furthermore, there was evidence that drug use in the Vernonia school, represented in the Supreme Court case* Vernonia v. Acton *(1995), was exaggerated, while use at the Pottawatomie school was underestimated. In their opinion, the court of appeals failed to take into account the unique dangers posed by illegal drugs.*

Theodore B. Olson et al., amicus curiae brief for the United States, *Board of Education of Independent School District No. 92 of Pottawatomie County v. Earls*, U.S. Supreme Court, December 2001.

The court [of appeals] emphasized that the policy here [*Pottawatomie v. Earls*] applies to non-athletic activities, and that the record of drug use here does not rise to the level of that in *Vernonia* [*v. Acton* (1995)]. Neither of those distinctions is of constitutional dimension.

The Policy's Application to Non-Athletes Does Not Render the Government Concern Any Less Significant

Here, as in *Vernonia*, "[t]hat the nature of the concern is important—indeed, perhaps compelling—can hardly be doubted." Deterring drug use by our Nation's schoolchildren is at least as important as enhancing efficient enforcement of the Nation's laws against the importation of drugs, which was the governmental concern in [*National Treasury Employees Union v.*] *Von Raab*, or deterring drug use by engineers and trainmen, which was the governmental concern in *Skinner* [*v. Railway Labor Executives' Association*]. The compelling nature of the government interest in deterring drug use in schools is alone sufficient to justify a drug-testing policy like the one at issue here.

Although it did not question the government's interest "in deterring drug use among students," the court of appeals suggested that that concern was less forceful here than in *Vernonia* on the ground that the risk of injury posed by drug use to student athletes is greater than to students engaged in the non-athletic activities covered by the policy in this case. To be sure, the linebacker faces a greater risk of serious injury if he takes the field under the influence of drugs than the drummer in the halftime band. But at the same time, the risk of injury to a student who is under the influence of drugs while playing golf, cross country, or volleyball (sports covered by the policy in Vernonia) is scarcely any greater than the risk of injury to a student who is drug-impaired while building a 15-foot-high human pyramid (as cheerleaders do), handling a 1500-pound steer (as FFA [Future Farmers of America] members do), or working with cutlery or sharp instruments (as FHA [Future Homemakers of America] members do). Moreover, students who participate in extracurricular activities, athletic or not, risk harm from drug use not just when they are competing, but at any time during the trips or overnight stays that they often take in connection with events.

Apart from the risk of injury to students who compete in [Oklahoma Secondary School Activities Association] OSSAA-governed events under the influence of drugs, the school district has another weighty interest in ensuring that students who participate in such events do so drug-free: when students step onto the auditorium stage for an academic "Quiz Bowl," the fairground for an FFA event, or the basketball court for Pom Pom, they do so on behalf of Tecumseh High. As the policy itself states, "[s]tudents who participate in [covered] activities are respected by the student body and are representing the school district and the community." Whatever the risk of injury from competing under the influence of drugs, the school district has a strong interest in seeing that those who represent Tecumseh High are not an "embarrassment" to the school and community, not to mention themselves.

In a similar vein, in *Vernonia*, this Court recognized that the "'role model' effect" of student athletes bolstered the government's interest in testing student athletes. So too here. Cheerleaders, for example, are just as likely to be role models as football players. And, as the record reflects, students who engage in non-athletic interscholastic competitions at Tecumseh High are held in "high esteem" by their fellow students. Indeed, the success of a non-athletic team, such as Tecumseh's academic team, which was state champion in 1997 and 1998, may actually call more attention to the members of that team than to members of sports teams with less success. The school district has a heightened interest in ensuring that those students do not use drugs.

At the same time, student athletes are no more entitled to protection from drugs, nor less entitled to legitimate concerns of privacy, than students who participate in non-athletic activities covered by the policy. Drawing a distinction between such students demeans both the volleyball player and the cheerleader by suggesting that one has a greater claim to privacy or is more deserving of protection from dangerous drugs than the other. There is, in short, no basis for adopting a constitutional dividing line based on whether a student chooses to participate in an interscholastic activity that is athletic, or one that is non-athletic.

The School District Was Not Required to Allow Drug Use at Tecumseh to Worsen Before Adopting Its Policy

Although the court of appeals recognized that "there was clearly some drug use at the Tecumseh schools," it doubted the "immediacy" of the school district's concern in addressing that problem because "the evidence of drug use among those subject to [its policy] is far from the 'epidemic' and 'immediate crisis' faced by the Vernonia schools." The court further held that "any district seeking to impose a random suspicionless drug-testing policy as a condition to participation in a school activity must demonstrate that there is some identifiable drug abuse problem among a sufficient number of those subject to the testing, such that testing that group will actually redress its drug problem." That analysis is flawed on several different levels and, if embraced by the Court, would seriously undercut local efforts to address the pervasive national problem posed by drugs.

First, as this Court's drug-testing cases confirm, the record in *Vernonia* by no means establishes the constitutional floor for justifying a random drug-testing program. Indeed, as the Court explained in *Vernonia*, the record in that case suggested an "immediate crisis of greater proportions than existed in

Skinner, where [the Court] upheld the Government's drug-testing program based on findings of drug use by railroad employees nationwide, without proof that a problem existed on the particular railroads whose employees were subject to the test. And of much greater proportions than existed in *Von Raab*, where there was no documented history of drug use by any Customs officials." Thus, far from raising the bar, *Vernonia* reaffirms that the government may establish a sufficiently important—and immediate—interest in adopting a random drug-testing program short of demonstrating that a societal drug problem has infiltrated a particular group of individuals.

Von Raab underscores the point. In that case, this Court specifically rejected the notion that, to justify its drug-testing program, the Customs Service was required to show drug use among the specific employees to be tested. Instead, the Court recognized that the government has a compelling interest in preventing drug use among the individuals to be tested; that there was no doubt that drug use was "one of the most serious problems confronting our society today"; and that there was "little reason to believe that American workplaces are immune from this pervasive social problem." In those circumstances, the Court stated, "[i]t is sufficient that the Government have a compelling interest in preventing an otherwise pervasive societal problem from spreading to the particular context." Without more, the government's undeniably compelling interest in deterring drug use among schoolchildren similarly justifies the school district's effort in this case to prevent drugs from spreading (further) to students subject to its policy. . . .

This case illustrates the problem with the court of appeals' approach. Illegal drug use remains pervasive among America's schoolchildren. Even if drug use were not yet a crisis in Oklahoma, the government would certainly have a compelling interest in preventing that crisis from occurring. But Oklahoma is not immune from the problem. It has one of the ten highest rates in the country of reported marijuana use among students ages 12 through 17. The record in this case further establishes that illicit drugs have reached Tecumseh schoolchildren, including those covered by the challenged policy. Against that backdrop, the school district was not required under this Court's precedents to wait until drug use became an epidemic before acting to save its children.

Evidence Shows That Tecumseh High Did Have a Drug Problem

Second, the evidence of drug use at Tecumseh High in fact paints a much more disturbing picture than the one portrayed by the court of appeals, including with respect to students covered by the policy. The court of appeals characterized the record of "actual drug use" in this case as "minimal," and respondents similarly claim an "almost complete absence of any drug problem at all in Tecumseh." That view cannot be reconciled with the record evidence, discussed above, that the school district adopted its policy in response to increasing calls for action by parents and concerned community members, and following reports by teachers, students, and parents of suspected or actual drug use among Tecumseh students (including those covered by the policy), and the discovery of drugs on school property.

At the same time that it underestimated the drug problem at Tecumseh High, the court of appeals arguably exaggerated the record of drug use in Vernonia. . . . Even accepting the court of appeals' characterization of the record in this case, the evidence of drug use in this case is not materially different from that in *Vernonia*. Moreover, in evaluating that evidence, it is important to recognize that, as this Court reiterated in *Vernonia*, a person impaired by drug use "will seldom display any outward 'signs detectable by the lay person or, in many cases, even the physician.'" Indeed, drug users go to considerable lengths to conceal their activities. In all likelihood, therefore, the prevalence of drug use at Tecumseh High is actually greater than what the record discloses.

Third, the court of appeals failed to accord appropriate deference to the judgments of local school officials as to the severity of the problem and need for action. The local board of education—in which parents of school children typically are represented—occupies a far better vantage point to gauge the threat posed by illegal drugs to their own schools and children than federal appellate judges. . . . The school district in this case adopted its drug-testing policy only after it had tried other anti-drug measures and had held community meetings to receive input on its proposed policy. The court of appeals erred in substituting its judgment for that of the school district as to the immediacy of the drug threat faced by Tecumseh students.

Fourth, as Judge Ebel [the dissenting judge of the court of appeals] explained, the court of appeals failed to take into account the unique dangers posed by illegal drugs and, in particular, the added difficulties in addressing a drug problem after drugs have become deeply entrenched in a community or its schools. In that regard, given the pervasive nature of the drug problem in schools nationwide and the incalculable damage inflicted on students and communities by illegal drugs, school administrators should have the same leeway to adopt the type of drug-testing policy at issue in this case to maintain a drug-free school as they do to adopt such a policy in an overdue effort to recreate one. Preventing drugs from "spreading" to a school is just as important as saving a drug-infested school, and much easier to do. Indeed, in *Vernonia*, this Court discussed the need to deter drug use, not just detect it. Schools across the Nation have a uniform interest in employing drug testing to deter drug use without regard to how much drug use is, in fact, detected. In short, schools have an interest in preventing irreparable damage before it occurs.

> "The [Court] dissenters found the lim-
> ited nature of the program 'perverse'
> because it 'targets for testing a student
> population least likely to be at risk from
> illicit drugs and their damaging
> effects.'"

The Student Drug-Testing Program Approved in *Earls* Is Not Fair

Craig M. Bradley

Craig M. Bradley is a professor of law at Indiana University. In the following viewpoint, he discusses the differences between the drug-testing program upheld by the Supreme Court in Pottawatomie v. Earls *(2002) and an earlier one that it approved in* Vernonia v. Acton *(1995). In the* Vernonia *case, only student athletes were tested, as they had been the leaders of a local drug culture. But, Bradley points out, in the* Earls *program, all participants in extracurricular activities were tested, and these students were the least likely to be taking illegal drugs. In his opinion, this was not fair and should not have been allowed by the Court. It would be fairer, he says, to test all students, yet such a program would have been struck down because it would provide no way to refuse testing short of being expelled from school. Bradley questions whether this decision may lead to the testing of adults when they choose certain activities, and concludes that it probably will not, since the Court emphasized that because schools are responsible for children, Fourth Amendment rights are different in schools than elsewhere.*

Craig M. Bradley, "Court Gives Drug-Testing an A," *Trial*, December 2002, pp. 56–57. Copyright © 2002 American Association for Justice, formerly Association of Trial Lawyers of America. Reproduced by permission.

In ... *Pottawatomie County v. Earls*, the Supreme Court upheld a school drug-testing program that was both broader and less justified than the program approved in 1995 in *Vernonia v. Acton*. Yet only Justice Ruth Bader Ginsburg switched her vote, concurring in the 6-3 *Vernonia* ruling and dissenting in the 5-4 *Earls* decision. In my view, Ginsburg has it right.

In *Vernonia*, as Ginsburg summarized in her *Earls* dissent, the Court had found that the testing program was limited to student athletes, that "drug use increase[d] the risk of sports-related injury, and that Vernonia's athletes were the leaders of an aggressive local drug culture that had reached epidemic proportions."

By contrast, the drug-testing program in *Earls*, instituted by the Tecumseh, Oklahoma, school district, extended to any student participating in extracurricular activities, including the academic team and Future Homemakers of America. There was no evidence that these students were the leaders of a drug culture in the community, and the drug problem in general was, according to the district superintendent, "not major." Because the school district had failed to demonstrate that there was a substantial problem among those subject to the testing program or that the testing program would redress such a problem, the Tenth Circuit had reversed the trial court and struck it down.

The Court's Decision

The Supreme Court, in an opinion by Justice Clarence Thomas, reinstated the program. The majority noted that for "special needs" searches, as opposed to searches for general law enforcement purposes, probable cause need not be shown. Rather, the program must be shown to be "reasonable." The Court made it clear that when considering what was reasonable, it would be most lenient in the school context, with a "hefty weight" on the school's side of the balance.

The Court simply wrote off as insignificant that the *Earls* program affected a much larger group of students than the one in *Vernonia* and that students on the academic team, for example, have not sacrificed personal privacy the way athletes in locker rooms have. The intrusion on privacy, the Court noted, was the same as in *Vernonia*, and the consequence of failing or refusing to take the test was only to limit the students' participation in extracurricular activities—not arrest, expulsion from school, or any other academic consequence.

While the evidence of a substantial drug problem was thin, the Court refused to second-guess the district court's finding that there was one. Then, the justices held that a "demonstrated problem of drug abuse is not in all cases necessary to the validity of a testing regime." However, the case relied on for that proposition, *National Treasury Employees Union v. Von Raab*, involved drug tests of Customs officials working in drug interdiction; those tests served a preventative need not present in *Earls*. Finally, although the potential for harm was not as critical an issue for the Future Homemakers of America as it was for student athletes, the "variety of health risks [of drug abuse] for children" overcame that fact.

Justice Stephen Breyer, whom one might have expected to defect from the *Vernonia* majority along with Ginsburg, wrote separately but joined the majority opinion. He emphasized the serious problem of drug abuse by teenagers, its deleterious effect on the public school system, and the fact that the program gives the "adolescent a nonthreatening reason" to resist peer pressure to use drugs. Finally, the program "avoids subjecting the entire school to testing. And it preserves an option for the conscientious objector. He can refuse testing while paying a price (nonparticipation) that is serious, but less severe than expulsion from the school."

By contrast, Ginsburg, writing for the four dissenters, found the limited nature of the program "perverse" because it "targets for testing a student population least likely to be at risk from illicit drugs and their damaging effects." Likewise, while student athletes might fairly be said to have agreed to a whole range of regulations, students participating in all other extracurricular activities should not be considered to have done the same. Such participation, she wrote, is "a key component of school life, essential in reality for students applying to college. . . . Students 'volunteer' for extracurricular pursuits in the same way they might volunteer for honors classes. They subject themselves to additional requirements, but they do so in order to take full advantage of the education offered them."

A Fairer System

Certainly, any school district that wants to pursue an aggressive drug-testing program is in a bit of a bind. A fairer system than the one in *Earls* would be to test *all* students, rather than only those who want to participate in extracurriculars. But such a program would have lost at least Breyer's crucial fifth vote and been struck down. Even though a program that singles out extracurricular participants is "perverse," the only consequence in the *Earls* program, as Breyer pointed out, was to be banned from participation. A school-wide program could be enforced only by expelling those who refused to be tested and could not, by definition, be avoided.

Ginsburg noted that the programs in *Vernonia* and *Earls* were alike in that they conditioned participation in certain activities by high school students on random urine testing. But there the similarities ended.

The Vernonia district sought to test a subpopulation of students distinguished by their reduced expectation of privacy, their special susceptibility to drug-related injury, and their heavy involvement with drug use. The Tecumseh district sought to test a much larger population associated with none of these factors.

If a school has a serious drug problem—and especially if athletes seem to be among the principal instigators of that problem—it does not seem unreasonable to condition participation in sports on drug testing. In contemporary America, where so many families are led by two working parents, parental supervision is reduced, making it reasonable for schools to assume a greater role in regulating student behavior.

But teachers or coaches cannot say, "Johnnie, if I ever hear that you've been smoking dope again, you're off the team," as parents can. Rather, a more objective system—drug testing—is the only way to control all but the most obviously intoxicated behavior. And the danger that intoxicated or hung-over athletes pose to themselves and others is real.

But tolerance for schools' efforts to prevent drug use by their students should not go so far as the school district did in *Earls*. Schoolchildren have Fourth Amendment rights, as the Court acknowledged. If those rights have any meaning at all, they should allow a student to participate in Future Homemakers of America without being required to provide a urine sample as a precondition of entering a pie in a baking contest. One hopes that, despite the Supreme Court's stamp of approval, most school districts will refrain from teaching this aspect of chemistry at the local high school.

Implications for the Future

Of greater concern might be what this decision augurs for those of us who have left our halcyon high school days behind. Should we expect soon to be required to submit to urinalysis when we wish to renew our driver's license or serve on the parks board? Probably not. In *Earls*, the Court emphasized that "Fourth Amendment rights . . . are different in public schools than elsewhere; the 'reasonableness' inquiry cannot disregard the schools' custodial and tutelary responsibility for children."

A 1997 case bolsters the Court's assertion that it will allow more intrusions into the privacy of schoolchildren than into that of other individuals. In *Chandler v. Miller* [1997], the

Court struck down a Georgia requirement that candidates for certain state offices submit to urinalysis within thirty days before qualifying for nomination or election. As the Court, in an 8-1 decision written by Ginsburg, declared:

> Georgia asserts no evidence of a drug problem among the state's elected officials, those officials typically do not perform high-risk, safety-sensitive tasks, and the required certification immediately aids no interdiction effort. The need revealed, in short, is symbolic, not "special" as that term draws meaning from our case law.

Much the same could have been said of the drug-testing program in *Earls*. Unfortunately, it was the dissent, rather than the majority, that said it.

"The decision elevates the myopic hysteria of a preposterous 'zero-tolerance' drug war over basic values such as respect and dignity for our nation's young people."

The Supreme Court's Ruling in *Earls* Teaches the Wrong Lesson to Students

Richard Glen Boire

Richard Glen Boire, an attorney, is senior fellow in law and policy at the Center for Cognitive Liberty & Ethics. In the following viewpoint, he argues that the drug-testing policy upheld by the Supreme Court in Pottawatomie v. Earls *will teach students that the Fourth Amendment is of mere historical interest and that dignity and respect are no longer valued. Furthermore, it will deter students who consider urine testing an invasion of privacy from participating in extracurricular activities, which will be harmful for society as well as for them, since such activities are the ones that build citizenship and prepare young people for leadership roles. Even in the case of students who do use drugs, he says, exclusion from extracurricular activities will do harm. In Boire's opinion, the war on drugs has not stopped drug use, but it has resulted in treating students and other citizens like criminal suspects and it therefore needs to be reconsidered.*

The Supreme Court's ruling on June 27, 2002, giving public school authorities the green light to conduct random, suspicious drug testing of all junior and senior high school students wishing to participate in extracurricular activities, teaches by example. The lesson, unfortunately, is that the Fourth Amendment has become a historical artifact, a quaint relic from bygone days when our country honored the "scrupulous protection of constitutional freedoms of the individual" [*West Virginia State Board of Education v. Barnette* (1943)].

The Court's ruling turns logic on its head, giving the insides of students' bodies less protection than the insides of their backpacks, the contents of their bodily fluids less protection than the contents of their telephone calls. The decision elevates the myopic hysteria of a preposterous "zero-tolerance" drug war over basic values such as respect and dignity for our nation's young people.

The Court's ruling treats America's teenage students like suspects. If a student seeks to participate in after-school activities, his or her urine can be taken and tested for any reason or for no reason at all. Gone are any requirements for individualized suspicion. Trust and respect have been replaced with a generalized distrust—an accusatory, authoritarian demand that students prove their "innocence" at the whim of the schoolmaster.

The Court majority reasoned that requiring students to yield up their urine for examination as a prerequisite to participating in extracurricular activities would serve as a deterrent to drug use. It reasoned that students who seek to join the debate team, write for the student newspaper, play in the marching band, or participate in any other after-school activities would be dissuaded from using drugs, knowing that their urine would be tested.

Participants in Extracurricular Activities
Are the Least Likely to Use Drugs

While some students may indeed be deterred from using drugs, the conventional wisdom (supported by empirical data) is that students who participate in extracurricular activities are some of the least likely to use drugs. Noting this, Justice Ruth Bader Ginsburg, whose dissenting opinion was joined by Justices John Paul Stevens, Sandra Day O'Connor, and David Souter, harshly condemned random testing of such students and describes it as "unreasonable, capricious and even perverse." Even when applied to students who do use drugs, the Court's decision merely makes matters worse.

The federal government has tried everything from threatening imprisonment to yanking student loans to spending hundreds of millions of dollars on "just say no" advertisements, and still some students continue to experiment with marijuana and other drugs. Like it or not, some students will use illegal drugs before graduating from high school, just as some students will have sex. Perhaps it's time to rethink the wisdom of declaring a "war on drugs" and adopt instead a realistic and effective strategy more akin to safe-sex education.

Ultimately, if a student does choose to experiment with an illegal drug (or a legal drug like alcohol), I suspect that many parents, like myself, would prefer that their child be taught the skills necessary to survive the experiment with as little harm as possible to self or others. The Drug Abuse Resistance Education (DARE) program—the nation's primary "drug" education curriculum—is taught by police officers, not drug experts, and is centered on intimidation and threats of criminal prosecution rather than on harm reduction. Random, suspicious urine testing fits the same tired mold.

Among the significant gaps in the majority's reasoning is its failure to consider the individual and social ramifications of deterring any student (whether or not they use drugs) from participating in after-school activities. Students who on principle prefer to keep their bodily fluid to themselves or who consider urine testing to be a gross invasion of privacy will be dissuaded from participating in after-school activities altogether. Similarly, students who do use drugs and who either test positive or forego the test for fear of what it might reveal will be banned from after-school activities and thus left to their own devices.

Policies That Discourage Participation in Extracurricular Activities Are Bad for Society

Extracurricular programs are valued for producing "well-rounded" students. Many adults look back on their extramural activities as some of the most educational, enriching, and formative experiences of their young lives. Extracurricular programs build citizenship, and for many universities, participation in after-school clubs and academic teams is a decisive admissions criterion. Whether or not students use drugs, it makes no sense to bar them from the very activities that build citizenship and help prepare them for leadership roles in the workforce, or help them get into college. In other words, a policy that deters students or bans them outright from participating in extracurricular activities isn't just bad for students, it's bad for society.

Aside from eviscerating the Fourth Amendment rights of the nation's 23 million public school students and imposing a punishment that harms society as much at it harms students, the decision foreshadows a constitutional dark age. When a young person is told to urinate in a cup within earshot of a school authority listening intently, and then ordered to turn over his or her urine for chemical examination, what "reasonable expectation of privacy" remains? When today's students graduate and walk out the schoolhouse gates, what will become of society's "reasonable expectation of privacy"?

Raised with the ever-present specter of coercion and control where urine testing is as common as standardized testing, today's students will have little if any privacy expectations when they reach adulthood. As a result, what society presently regards as a "reasonable expectation of privacy" will be considerably watered down within a single generation. Rivers of urine will have eroded the Fourth Amendment—our nation's strictest restraint on the overreaching and strong-arm tendencies of some government police agents. Justice Ginsburg and the three other justices who joined her dissenting opinion aptly state "that [schools] are educating the young for citizenship is reason for scrupulous protection of Constitutional freedoms of the individual, if we are not to strangle the free mind at its source and teach youth to discount important principles of our government as mere platitudes."

The U.S. government has just allocated another $19 billion to fight the so-called war on drugs, yet all we really have to show for it is a tattered Constitution and the largest prison population in the history of the world. Fellow U.S. citizens have been constructed as "the enemy" simply because they'd rather have a puff of marijuana than a shot of bourbon. And that is perhaps the greatest tragedy of the Court's ruling. The decision not only victimizes our children, but it makes them the enemy. Being a public school student is now synonymous with being a criminal suspect or a prisoner.

The values of trust and respect have been chased from the schoolyards and replaced with baseless suspicion and omnipresent policing. The lesson for U.S. students as they stand in line with urine bottles in hand is that the Fourth Amendment's guarantee is now a broken promise, yesterday's dusty trophy, and worthy only of lip service. The lesson for the rest of us is that the so-called war on drugs desperately needs rethinking.

CHAPTER 4

The Federal Government Can Prohibit Medical Use of Marijuana

Case Overview

Gonzales v. Raich (2005)

Angel Raich suffers from serious medical problems. She has an inoperable brain tumor, life-threatening wasting syndrome, seizure disorder, nausea, and a number of other conditions. She was in a wheelchair for several years. She has severe chronic pain, and, because of extreme allergic reactions, she cannot use standard medications to control it. Therefore, her doctor suggested that she try cannabis (marijuana), which she was able to do because California, where she lives, is one of nine states where medical use of marijuana is legal. This enabled her to get out of the wheelchair and helped her to eat. If she could no longer use it, she would risk malnutrition and ultimately starvation. "Without cannabis my life would be a death sentence," she says.

Her doctor agrees. "Angel has no reasonable legal alternative to cannabis for the effective treatment or alleviation of her medical conditions or symptoms associated with the medical conditions," he says, "because she has tried essentially all other legal alternatives to cannabis and the alternatives have been ineffective or result in intolerable side effects. Angel will suffer imminent harm without access to cannabis."

Angel is physically unable to grow marijuana, so two anonymous caregivers cultivate and process it for her at no charge. Another medical user, Diane Monson, does grow her own. In 2002, federal Drug Enforcement Administration (DEA) agents raided Monson's home and destroyed her marijuana plants, despite the fact that she was legally entitled to them under California law. The two women, Monson and Raich, brought suit against the federal government, claiming that enforcing the Controlled Substances Act (CSA) against them would violate the commerce clause of the Constitution, the due process clause of the Fifth Amendment, the Ninth and Tenth Amendments, and the doctrine of medical necessity. The District Court refused to issue an order preventing its enforcement, but the Court of Appeals did so, acknowledging that they had "demonstrated a strong likelihood of success on their claim that, as applied to them, the CSA is an unconstitutional exercise of Congress' commerce clause authority." The government then took its case to the U.S. Supreme Court.

Although this case appears to be about medical use of marijuana, it actually had a quite different focus. The question to be decided was not whether allowing such use is justifiable, but whether the federal government has the authority to make laws that preempt state laws. Under the commerce clause of the Constitution, it does have this power when interstate commerce is involved. Angel Raich's attorney argued that locally grown marijuana that is never sold has nothing to do with interstate commerce—or for that matter, with commerce at all. The government argued that it does, because it affects the total amount of marijuana that exists and therefore has an impact on the national drug market.

In a six to three decision, the Court ruled that the CSA applies even in states where medical use of marijuana is allowed. Supporters of the government's position were concerned about more than the issue of illegal drugs. They pointed out that many federal laws, including civil rights and environmental protection laws, are based on the commerce clause and that if it were not interpreted broadly, the validity of those other laws might be in doubt. Opponents, on the other hand, viewed federal regulation of purely local conduct as a violation of the Tenth Amendment, which reserves to the states all powers not specifically given to the federal government by the Constitution. They feared that if the federal government had the power to control activity occurring entirely within a state's own borders, it might start regulating all sorts of things that have no true relationship to interaction between the states. Two of the justices wrote dissenting opinions; in his, Justice Clarence Thomas said, "If the majority is to be taken seriously, the Federal Government may now regulate quilting bees, clothes drives, and potluck suppers throughout the fifty States."

That some patients would suffer and might even die without medical marijuana was not relevant to the Court's decision, as Congress had stated that marijuana has no legitimate medical use. The plaintiffs' argument that the due process clause and the doctrine of medical necessity were violated by enforcement of the CSA had not been dealt with by the lower court and the case was remanded to it for those issues to be considered. The Court also suggested that the best remedy for patients would be to seek a change in the law through the democratic process.

When the case went back to the lower court and Angel Raich pursued her claim of medical necessity, the Ninth Circuit Court of Appeals ruled against her. However, it indicated that if she were ever arrested and prosecuted for marijuana use, she could use as a defense the principle that when a person is forced to choose between her life and disobeying a criminal law, she may not be punished for preserving her life.

*"The exemption for cultivation by pa-
tients and caregivers can only increase
the supply of marijuana in the California market. The likelihood that all such
production will promptly terminate
when patients recover . . . seems remote."*

The Court's Decision:
Congress Has the Power
to Regulate Drugs Despite
State Laws

John Paul Stevens

*John Paul Stevens is, as of 2008, the oldest and longest-serving
member of the Supreme Court, and is generally considered to be
the leader of its liberal faction. In the following majority opinion
in* Gonzales v. Raich, *he states that the use of marijuana for
medical purposes, even where legal under state law, is a violation of the Controlled Substances Act and cannot be exempted
under federal law. Medical users had claimed that Congress did
not have the authority to ban it under the commerce clause of
the Constitution, which gives the federal government the power
to regulate interstate commerce. The growing of small amounts
of marijuana by patients and their caregivers is not commerce,
they argued. Justice Stevens maintains that it does affect commerce even though the marijuana is grown locally and is not
sold, because it has an impact on availability and thus affects
the national market. One class of illegal drugs cannot be sepa-*

John Paul Stevens, majority opinion, *Gonzales v. Raich*, U.S. Supreme Court, June 6,
2005.

rated from the regulatory scheme as whole, he says, and in any case, no marijuana can be treated as a separate class on the grounds of its being used for medicinal purposes, since the government has ruled that it has no legitimate medical use. He suggests that patients who need it should aim for a change in the law through the democratic process.

Shortly after taking office in 1969, President [Richard] Nixon declared a national "war on drugs." As the first campaign of that war, Congress set out to enact legislation that would consolidate various drug laws on the books into a comprehensive statute, provide meaningful regulation over legitimate sources of drugs to prevent diversion into illegal channels, and strengthen law enforcement tools against the traffic in illicit drugs. That effort culminated in the passage of the Comprehensive Drug Abuse Prevention and Control Act of 1970. . . .

Title II of that Act, the CSA [Controlled Substances Act] repealed most of the earlier anti-drug laws in favor of a comprehensive regime to combat the international and interstate traffic in illicit drugs. The main objectives of the CSA were to conquer drug abuse and to control the legitimate and illegitimate traffic in controlled substances. Congress was particularly concerned with the need to prevent the diversion of drugs from legitimate to illicit channels. . . .

In enacting the CSA, Congress classified marijuana as a Schedule I drug. . . . Schedule I drugs are categorized as such because of their high potential for abuse, lack of any accepted medical use, and absence of any accepted safety for use in medically supervised treatment. . . . By classifying marijuana as a Schedule I drug, as opposed to listing it on a lesser schedule, the manufacture, distribution, or possession of marijuana became a criminal offense, with the sole exception being use of the drug as part of a Food and Drug Administration pre-approved research study. . . .

Respondents in this case do not dispute that passage of the CSA, as part of the Comprehensive Drug Abuse Prevention and Control Act, was well within Congress' commerce power. Nor do they contend that any provision or section of the CSA amounts to an unconstitutional exercise of congressional authority. Rather, respondents' challenge is actually quite limited; they argue that the CSA's categorical prohibition of the manufacture and possession of marijuana as applied to the intrastate manufacture and possession of marijuana for medical purposes pursuant to California law exceeds Congress' authority under the Commerce Clause.

In assessing the validity of congressional regulation, none of our Commerce Clause cases can be viewed in isolation. As charted in considerable detail in *United States v. Lopez* [(1995)], our understanding of the reach of the Commerce Clause, as well as Congress' assertion of authority thereunder, has evolved over time. The Commerce Clause emerged as the Framers' response to the central problem giving rise to the Constitution itself: the absence of any federal commerce power under the Articles of Confederation. For the first century of our history, the primary use of the Clause was to preclude the kind of discriminatory state legislation that had once been permissible. Then, in response to rapid industrial development and an increasingly interdependent national economy, Congress "ushered in a new era of federal regulation under the commerce power," beginning with the enactment of the Interstate Commerce Act in 1887, and the Sherman Antitrust Act in 1890. . . .

The Precedent Set by *Wickard* Is Important

Our case law firmly establishes Congress' power to regulate purely local activities that are part of an economic "class of activities" that have a substantial effect on interstate commerce. As we stated in *Wickard v. Filburn* (1942), "even if appellee's activity be local and though it may not be regarded as commerce, it may still, whatever its nature, be reached by Congress if it exerts a substantial economic effect on interstate commerce." We have never required Congress to legislate with scientific exactitude. When Congress decides that the "'total incidence'" of a practice poses a threat to a national market, it may regulate the entire class. . . .

Our decision in *Wickard*, is of particular relevance. In *Wickard*, we upheld the application of regulations promulgated under the Agricultural Adjustment Act of 1938, which were designed to control the volume of wheat moving in interstate and foreign commerce in order to avoid surpluses and consequent abnormally low prices. The regulations established an allotment of 11.1 acres for Filburn's 1941 wheat crop, but he sowed 23 acres, intending to use the excess by consuming it on his own farm. Filburn argued that even though we had sustained Congress' power to regulate the production of goods for commerce, that power did not authorize "federal regulation [of] production not intended in any part for commerce but wholly for consumption on the farm." Justice [Robert] Jackson's opinion for a unanimous Court rejected this submission. He wrote:

> The effect of the statute before us is to restrict the amount which may be produced for market and the extent as well to which one may forestall resort to the market by producing to meet his own needs. That appellee's own contribution to the demand for wheat may be trivial by itself is not enough to remove him from the scope of federal regulation where, as here, his contribution, taken together with that of many others similarly situated, is far from trivial.

Wickard thus establishes that Congress can regulate purely intrastate activity that is not itself "commercial," that it is not produced for sale, if it concludes that failure to regulate that class of activity would undercut the regulation of the interstate market in that commodity.

The similarities between this case and *Wickard* are striking. Like the farmer in *Wickard*, respondents are cultivating for home consumption a fungible commodity for which there is an established, albeit illegal, interstate market. Just as the Agricultural Adjustment Act was designed "to control the volume [of wheat] moving in interstate and foreign commerce in order to avoid surpluses . . ." and consequently control the market price, a primary purpose of the CSA is to control the supply and demand of controlled substances in both lawful and unlawful drug markets. In *Wickard*, we had no difficulty concluding that Congress had a rational basis for believing that, when viewed in the aggregate, leaving home-consumed wheat outside the regulatory scheme would have a substantial influence on price and market conditions. Here too, Congress had a rational basis for concluding that leaving home-consumed marijuana outside federal control would similarly affect price and market conditions.

More concretely, one concern prompting inclusion of wheat grown for home consumption in the 1938 Act was that rising market prices could draw such wheat into the interstate market, resulting in lower market prices. The parallel concern making it appropriate to include marijuana grown for home consumption in the CSA is the likelihood that the high demand in the interstate market will draw such marijuana into that market. While the diversion of homegrown wheat tended to frustrate the federal interest in stabilizing prices by regulating the volume of commercial transactions in the interstate market, the diversion of homegrown marijuana tends to frustrate the federal interest in eliminating commercial transactions in the interstate market in their entirety. In both cases, the regulation is squarely within Congress' commerce power because production of the commodity meant for home consumption, be it wheat or marijuana, has a substantial effect on supply and demand in the national market for that commodity.

Nonetheless, respondents suggest that *Wickard* differs from this case in three respects: (1) the Agricultural Adjustment Act, unlike the CSA, exempted small farming operations; (2)

Wickard involved a "quintessential economic activity"—a commercial farm—whereas respondents do not sell marijuana, and (3) the *Wickard* record made it clear that the aggregate production of wheat for use on farms had a significant impact on market prices. Those differences, though factually accurate, do not diminish the precedential force of this Court's reasoning.

The fact that Wickard's own impact on the market was "trivial by itself" was not a sufficient reason for removing him from the scope of federal regulation. That the Secretary of Agriculture elected to exempt even smaller farms from regulation does not speak to his power to regulate all those whose aggregated production was significant, nor did that fact play any role in the Court's analysis. Moreover, even though Wickard was indeed a commercial farmer, the activity he was engaged in—the cultivation of wheat for home consumption—was not treated by the Court as part of his commercial farming operation. And while it is true that the record in the *Wickard* case itself established the causal connection between the production for local use and the national market, we have before us findings by Congress to the same effect.

Is There a Rational Basis for the Claim That Medical Marijuana Affects Interstate Commerce?

Findings in the introductory sections of the CSA explain why Congress deemed it appropriate to encompass local activities within the scope of the CSA. Submissions of the parties and the numerous *amici* [friends of the court briefs] all seem to agree that the national and international market for marijuana has dimensions that are fully comparable to those defining the class of activities regulated by the Secretary pursuant to the 1938 statute [Agricultural Adjustment Act]. Respondents nonetheless insist that the CSA cannot be constitutionally applied to their activities, because Congress did not make a specific finding that the intrastate cultivation and possession of marijuana for medical purposes based on the recommendation of a physician would substantially affect the larger interstate marijuana market. Be that as it may, we have never required Congress to make particularized findings in order to legislate, absent a special concern such as the protection of free speech. While congressional findings are certainly helpful in reviewing the substance of a congressional statutory scheme, particularly when the connection to commerce is not self-evident, and while we will consider congressional findings in our analysis when they are available, the absence of particularized findings does not call into question Congress' authority to legislate.

In assessing the scope of Congress' authority under the Commerce Clause, we stress that the task before us is a modest one. We need not determine whether respondents' activities, taken in the aggregate, substantially affect interstate commerce in fact, but only whether a "rational basis" exists for so concluding. Given the enforcement difficulties that attend distinguishing between marijuana cultivated locally and marijuana grown elsewhere, and concerns about diversion into illicit channels, we have no difficulty concluding that Congress had a rational basis for believing that failure to regulate the intrastate manufacture and possession of marijuana would leave a gaping hole in the CSA. Thus, as in *Wickard*, when it enacted comprehensive legislation to regulate the interstate market in a fungible commodity, Congress was acting well within its authority to "make all Laws which shall be necessary and proper" to "regulate Commerce . . . among the several States." That the regulation ensnares some purely intrastate activity is of no moment. As we have done many times before, we refuse to excise individual components of that larger scheme. . . .

Use of Marijuana for Medical Purposes Does Not Exempt It from the CSA

The Court [of Appeals] characterized [users of medical marijuana] as "different in kind from drug trafficking." The differences between the members of a class so defined and the principal traffickers in Schedule I substances might be sufficient to justify a policy decision exempting the narrower class from the coverage of the CSA. The question, however, is whether Congress' contrary policy judgment, *i.e.*, its decision to include this narrower "class of activities" within the larger regulatory scheme, was constitutionally deficient. We have no difficulty concluding that Congress acted rationally in determining that none of the characteristics making up the purported class, whether viewed individually or in the aggregate, compelled an exemption from the CSA; rather, the subdivided class of activities defined by the Court of Appeals was an essential part of the larger regulatory scheme.

First, the fact that marijuana is used "for personal medical purposes on the advice of a physician" cannot itself serve as a distinguishing factor. The CSA designates marijuana as contraband for any purpose; in fact, by characterizing marijuana as a Schedule I drug, Congress expressly found that the drug has no acceptable medical uses. Moreover, the CSA is a comprehensive regulatory regime specifically designed to regulate which controlled substances can be utilized for medicinal purposes, and in what manner. Indeed, most of the substances classified in the CSA "have a useful and legitimate medical purpose." Thus, even if respondents are correct that marijuana does have accepted medical uses and thus should be redesignated as a lesser schedule drug, the CSA would still impose controls beyond what is required by California law. The CSA requires manufacturers, physicians, pharmacies, and other handlers of controlled substances to comply with statutory and regulatory provisions mandating registration with the DEA, compliance with specific production quotas, security controls to guard against diversion, record keeping and reporting obligations, and prescription requirements. Furthermore, the dispensing of new drugs, even when doctors approve their use, must await federal approval. Accordingly, the mere fact that marijuana—like virtually every other controlled substance regulated by the CSA—is used for medicinal purposes cannot possibly serve to distinguish it from the core activities regulated by the CSA.

Nor can it serve as an "objective marke[r]" or "objective facto[r]" to arbitrarily narrow the relevant class as the dissenters suggest. More fundamentally, if, as the principal dissent contends, the personal cultivation, possession, and use of marijuana for medicinal purposes is beyond the "'outer limits' of Congress' Commerce Clause authority," it must also be true that such personal use of marijuana (or any other homegrown drug) for recreational purposes is also beyond those "'outer limits,'" whether or not a State elects to authorize or even regulate such use. . . . That is, the dissenters' rationale logically extends to place *any* federal regulation (including quality, prescription, or quantity controls) of *any* locally cultivated and possessed controlled substance for *any* purpose beyond the "'outer limits'" of Congress' Commerce Clause authority. One need not have a degree in economics to understand why a nationwide exemption for the vast quantity of marijuana (or other drugs) locally cultivated for personal use (which presumably would include use by friends, neighbors, and family members) may have a substantial impact on the interstate market for this extraordinarily popular substance. The congressional judgment that an exemption for such a significant segment of the total market would undermine the orderly enforcement of the entire regulatory scheme is entitled to a strong presumption of validity. Indeed, that judgment is not only rational, but "visible to the naked eye," under any commonsense appraisal of the probable consequences of such an open-ended exemption.

Federal Law Prevails over State Law

Second, limiting the activity to marijuana possession and cultivation "in accordance with state law" cannot serve to place respondents' activities beyond congressional reach. The Supremacy Clause unambiguously provides that if there is any conflict between federal and state law, federal law shall prevail. It is beyond peradventure that federal power over commerce is "'superior to that of the States to provide for the welfare or necessities of their inhabitants,'" however legitimate or dire those necessities may be. . . .

That the California exemptions will have a significant impact on both the supply and demand sides of the market for marijuana is not just "plausible" as the principal dissent concedes, it is readily apparent. The exemption for physicians provides them with an economic incentive to grant their patients permission to use the drug. In contrast to most prescriptions for legal drugs, which limit the dosage and duration of the usage, under California law the doctor's permission to recommend marijuana use is open-ended. The authority to grant permission whenever the doctor determines that a patient is afflicted with "any other illness for which marijuana provides relief," is broad enough to allow even the most scrupulous doctor to conclude that some recreational uses would be therapeutic. And our cases have taught us that there are some unscrupulous physicians who overprescribe when it is sufficiently profitable to do so.

The exemption for cultivation by patients and caregivers can only increase the supply of marijuana in the California market. The likelihood that all such production will promptly terminate when patients recover or will precisely match the patients' medical needs during their convalescence seems remote, whereas the danger that excesses will satisfy some of the admittedly enormous demand for recreational use seems obvious. Moreover, that the national and international narcotics trade has thrived in the face of vigorous criminal enforcement efforts suggests that no small number of unscrupulous people will make use of the California exemptions to serve their commercial ends whenever it is feasible to do so. Taking into account the fact that California is only one of at least nine States to have authorized the medical use of marijuana, a fact Justice [Sandra Day] O'Connor's dissent conveniently disregards in arguing that the demonstrated effect on commerce while admittedly "plausible" is ultimately "unsubstantiated," Congress could have rationally concluded that the aggregate impact on the national market of all the transactions exempted from federal supervision is unquestionably substantial.

So, from the "separate and distinct" class of activities identified by the Court of Appeals (and adopted by the dissenters), we are left with "the intrastate, noncommercial cultivation, possession and use of marijuana." Thus the case for the exemption comes down to the claim that a locally cultivated product that is used domestically rather than sold on the open market is not subject to federal regulation. Given the findings in the CSA and the undisputed magnitude of the commercial market for marijuana, our decisions in *Wickard v. Filburn* and the later cases endorsing its reasoning foreclose that claim.

Respondents also raise a substantive due process claim and seek to avail themselves of the medical necessity defense. These theories of relief were set forth in their complaint but were not reached by the Court of Appeals. We therefore do not address the question whether judicial relief is available to respondents on these alternative bases. We do note, however, the presence of another avenue of relief. As the Solicitor General confirmed during oral argument, the statute authorizes procedures for the reclassification of Schedule I drugs. But perhaps even more important than these legal avenues is the democratic process, in which the voices of voters allied with these respondents may one day be heard in the halls of Congress. Under the present state of the law, however, the judgment of the Court of Appeals must be vacated.

> "Whatever the wisdom of California's experiment with medical marijuana, the federalism principles that have driven our Commerce Clause cases require that room for experiment be protected in this case."

Dissenting Opinion: The Constitution Leaves Protection of Citizens' Health up to the States

Sandra Day O'Connor

Sandra Day O'Connor was the first woman justice of the Supreme Court, where she served from 1981 until her retirement in 2006. She was considered a swing vote between its conservative and liberal factions. In the following dissenting opinion in Gonzales v. Raich, *she argues that one of the virtues of America's federal system of government is that individual states can experiment without risk to the rest of the country, and that states have always had the authority to protect the health and welfare of their citizens. The Court's ruling, she says, puts an end to experiments with legalizing the use of marijuana for medicinal purposes without evidence that personal cultivation and possession of medical marijuana has any effect on interstate commerce—which is the only basis on which Congress has the right to regulate it. In her opinion, the precedents on which the Court relied are not applicable because those cases were not comparable to*

Sandra Day O'Connor, dissenting opinion, *Gonzales v. Raich*, U.S. Supreme Court, June 6, 2005.

this one. Although she herself does not favor state laws permitting medical use of marijuana, she believes that the Constitution's commerce clause does not allow the federal government to overrule them.

We enforce the "outer limits" of Congress' Commerce Clause authority not for their own sake, but to protect historic spheres of state sovereignty from excessive federal encroachment and thereby to maintain the distribution of power fundamental to our federalist system of government. One of federalism's chief virtues, of course, is that it promotes innovation by allowing for the possibility that "a single courageous State may, if its citizens choose, serve as a laboratory, and try novel social and economic experiments without risk to the rest of the country."

This case exemplifies the role of States as laboratories. The States' core police powers have always included authority to define criminal law and to protect the health, safety, and welfare of their citizens. Exercising those powers, California (by ballot initiative and then by legislative codification) has come to its own conclusion about the difficult and sensitive question of whether marijuana should be available to relieve severe pain and suffering. Today the Court sanctions an application of the federal Controlled Substances Act that extinguishes that experiment, without any proof that the personal cultivation, possession, and use of marijuana for medicinal purposes, if economic activity in the first place, has a substantial effect on interstate commerce and is therefore an appropriate subject of federal regulation. In so doing, the Court announces a rule that gives Congress a perverse incentive to legislate broadly pursuant to the Commerce Clause—nestling questionable assertions of its authority into comprehensive regulatory schemes—rather than with precision. That rule and the result it produces in this case are irreconcilable with our decisions in [*United States v.*] *Lopez* [(1995)], and *United States v. Morrison* (2000). Accordingly I dissent.

In *Lopez*, we considered the constitutionality of the Gun-Free School Zones Act of 1990, which made it a federal offense "for any individual knowingly to possess a firearm . . . at a place the individual knows or has reasonable cause to believe is a school zone." . . .

Our decision about whether gun possession in school zones substantially affected interstate commerce turned on four considerations. First, we observed that our "substantial effects" cases generally have upheld federal regulation of economic activity that affected interstate commerce, but that [the Gun-Free School Zones Act] was a criminal statute having "nothing to do with 'commerce' or any sort of economic enterprise." . . .

Second, we noted that the statute contained no express jurisdictional requirement establishing its connection to interstate commerce.

Third, we found telling the absence of legislative findings about the regulated conduct's impact on interstate commerce. We explained that while express legislative findings are neither required nor, when provided, dispositive, findings "enable us to evaluate the legislative judgment that the activity in question substantially affect[s] interstate commerce, even though no such substantial effect [is] visible to the naked eye." Finally, we rejected as too attenuated the Government's argument that firearm possession in school zones could result in violent crime which in turn could adversely affect the national economy. The Constitution, we said, does not tolerate reasoning that would "convert congressional authority under the Commerce Clause to a general police power of the sort retained by the States." Later in *Morrison*, we relied on the same four considerations to hold that the Violence Against Women Act of 1994, exceeded Congress' authority under the Commerce Clause.

In my view, the case before us is materially indistinguishable from *Lopez* and *Morrison* when the same considerations are taken into account.

The Court's Decision Removes Limits on the Commerce Clause

What is the relevant conduct subject to Commerce Clause analysis in this case? The Court takes its cues from Congress, applying the above considerations to the activity regulated by the Controlled Substances Act (CSA) in general. The Court's decision rests on two facts about the CSA: (1) Congress chose to enact a single statute providing a comprehensive prohibition on the production, distribution, and possession of all controlled substances, and (2) Congress did not distinguish between various forms of intrastate noncommercial cultivation, possession, and use of marijuana. Today's decision suggests that the federal regulation of local activity is immune to Commerce Clause challenge because Congress chose to act with an ambitious, all-encompassing statute, rather than piecemeal. In my view, allowing Congress to set the terms of the constitutional debate in this way, i.e., by packaging regulation of local activity in broader schemes, is tantamount to removing meaningful limits on the Commerce Clause.

The Court's principal means of distinguishing *Lopez* from this case is to observe that the Gun-Free School Zones Act of 1990 was a "brief, single-subject statute," whereas the CSA is "a lengthy and detailed statute creating a comprehensive framework for regulating the production, distribution, and possession of five classes of 'controlled substances.'" Thus, according to the Court, it was possible in *Lopez* to evaluate in isolation the constitutionality of criminalizing local activity (there gun possession in school zones), whereas the local activity that the CSA targets (in this case cultivation and possession of marijuana for personal medicinal use) cannot be separated from the general drug control scheme of which it is a part.

Today's decision allows Congress to regulate intrastate activity without check so long as there is some implication by legislative design that regulating intrastate activity is essential (and the Court appears to equate "essential" with "necessary") to the interstate regulatory scheme. Seizing upon our language in *Lopez* that the statute prohibiting gun possession in school zones was "not an essential part of a larger regulation of economic activity, in which the regulatory scheme could be undercut unless the intrastate activity were regulated," the Court appears to reason that the placement of local activity in a comprehensive scheme confirms that it is essential to that scheme. If the Court is right, then *Lopez* stands for nothing more than a drafting guide: Congress should have described the relevant crime as "transfer or possession of a firearm anywhere in the nation"—thus including commercial and noncommercial activity and clearly encompassing some activity with assuredly substantial effect on interstate commerce. Had it done so, the majority hints, we would have sustained its authority to regulate possession of firearms in school zones. . . .

Lopez and *Morrison* did not indicate that the constitutionality of federal regulation depends on superficial and formalistic distinctions. Likewise, I did not understand our discussion of the role of courts in enforcing outer limits of the Commerce Clause for the sake of maintaining the federalist balance our Constitution requires, as a signal to Congress to enact legislation that is more extensive and more intrusive into the domain of state power. If the Court always defers to Congress as it does today, little may be left to the notion of enumerated powers.

A Line Should Be Drawn Between Local and National Economics

The hard work for courts, then, is to identify objective markers for confining the analysis in Commerce Clause cases. Here, respondents challenge the constitutionality of the CSA as applied to them and those similarly situated. I agree with the Court that we must look beyond respondents' own activities. Otherwise, individual litigants could always exempt themselves from Commerce Clause regulation merely by pointing to the obvious—that their personal activities do not have a substantial effect on interstate commerce. The task is to identify a mode of analysis that allows Congress to regulate more than nothing (by declining to reduce each case to its litigants) and less than everything (by declining to let Congress set the terms of analysis). The analysis may not be the same in every case, for it depends on the regulatory scheme at issue and the federalism concerns implicated.

A number of objective markers are available to confine the scope of constitutional review here. Both federal and state legislation—including the CSA itself, the California Compassionate Use Act, and other state medical marijuana legislation—recognize that medical and nonmedical (*i.e.*, recreational) uses of drugs are realistically distinct and can be segregated, and regulate them differently. Respondents challenge only the application of the CSA to medicinal use of marijuana. Moreover, because fundamental structural concerns about dual sovereignty animate our Commerce Clause cases, it is relevant that this case involves the interplay of federal and state regulation in areas of criminal law and social policy, where "States lay claim by right of history and expertise." California, like other States, has drawn on its reserved powers to distinguish the regulation of medicinal marijuana. To ascertain whether Congress' encroachment is constitutionally justified in this case, then, I would focus here on the personal cultivation, possession, and use of marijuana for medicinal purposes.

Having thus defined the relevant conduct, we must determine whether, under our precedents, the conduct is economic and in the aggregate substantially affects interstate commerce. Even if intrastate cultivation and possession of marijuana for one's own medicinal use can properly be characterized as economic, and I question whether it can, it has not been shown that such activity substantially affects interstate commerce.

Similarly, it is neither self-evident nor demonstrated that regulating such activity is necessary to the interstate drug control scheme.

The Court's definition of economic activity is breathtaking. It defines as economic any activity involving the production, distribution, and consumption of commodities. And it appears to reason that when an interstate market for a commodity exists, regulating the intrastate manufacture or possession of that commodity is constitutional either because that intrastate activity is itself economic or because regulating it is a rational part of regulating its market. Putting to one side the problem endemic to the Court's opinion—the shift in focus from the activity at issue in this case to the entirety of what the CSA regulates—the Court's definition of economic activity for purposes of Commerce Clause jurisprudence threatens to sweep all of productive human activity into federal regulatory reach.

The Court uses a dictionary definition of economics to skirt the real problem of drawing a meaningful line between "what is national and what is local." It will not do to say that Congress may regulate noncommercial activity simply because it may have an effect on the demand for commercial goods, or because the noncommercial endeavor can, in some sense, substitute for commercial activity. Most commercial goods or services have some sort of privately producible analogue. Home care substitutes for day care. Charades games substitute for movie tickets. Backyard or windowsill gardening substitutes for going to the supermarket. To draw the line wherever private activity affects the demand for market goods is to draw no line at all, and to declare everything economic. We have already rejected the result that would follow—a federal police power. . . .

The homegrown cultivation and personal possession and use of marijuana for medicinal purposes has no apparent commercial character. Everyone agrees that the marijuana at issue in this case was never in the stream of commerce, and neither were the supplies for growing it. (Marijuana is highly unusual among the substances subject to the CSA in that it can be cultivated without any materials that have traveled in interstate commerce.) *Lopez* makes clear that possession is not itself commercial activity. And respondents have not come into possession by means of any commercial transaction; they have simply grown, in their own homes, marijuana for their own use, without acquiring, buying, selling, or bartering a thing of value.

This Case Is Different from the *Wickard* Case

The Court suggests that *Wickard*, which we have identified as "perhaps the most far-reaching example of Commerce Clause authority over intrastate activity," established federal regulatory power over any home consumption of a commodity for which a national market exists. I disagree. *Wickard* involved a challenge to the Agricultural Adjustment Act of 1938 (AAA), which directed the Secretary of Agriculture to set national quotas on wheat production and penalties for excess production. The AAA itself confirmed that Congress made an explicit choice not to reach—and thus the Court could not possibly have approved of federal control over—small-scale noncommercial wheat farming. In contrast to the CSA's limitless assertion of power, Congress provided an exemption within the AAA for small producers. When Filburn planted the wheat at issue in *Wickard*, the statute exempted plantings less than 200 bushels (about six tons), and when he harvested his wheat, it exempted plantings less than six acres. *Wickard*, did not extend Commerce Clause authority to something as modest as the home cook's herb garden. This is not to say that Congress may never regulate small quantities of commodities possessed or produced for personal use, or to deny that it sometimes needs to enact a zero-tolerance regime for such commodities.

It is merely to say that *Wickard* did not hold or imply that small-scale production of commodities is always economic and automatically within Congress' reach.

Even assuming that economic activity is at issue in this case, the Government has made no showing in fact that the possession and use of homegrown marijuana for medical purposes in California or elsewhere has a substantial effect on interstate commerce. Similarly, the Government has not shown that regulating such activity is necessary to an interstate regulatory scheme. Whatever the specific theory of "substantial effects" at issue (*i.e.*, whether the activity substantially affects interstate commerce, whether its regulation is necessary to an interstate regulatory scheme, or both), a concern for dual sovereignty requires that Congress' excursion into the traditional domain of States be justified.

There is simply no evidence that homegrown medicinal marijuana users constitute, in the aggregate, a sizable enough class to have a discernable, let alone substantial, impact on the national illicit drug market—or otherwise to threaten the CSA regime. Explicit evidence is helpful when substantial effect is not "visible to the naked eye." And here, in part because common sense suggests that medical marijuana users may be limited in number and that California's Compassionate Use Act and similar state legislation may well isolate activities relating to medicinal marijuana from the illicit market, the effect of those activities on interstate drug traffic is not self-evidently substantial.

In this regard again, this case is readily distinguishable from *Wickard*. To decide whether the Secretary could regulate local wheat farming, the Court looked to "the actual effects of the activity in question upon interstate commerce." Critically, the Court was able to consider "actual effects" because the parties had "stipulated a summary of the economics of the wheat industry." . . .

The Court recognizes that "the record in the *Wickard* case itself established the causal connection between the production for local use and the national market" and argues that "we have before us findings by Congress *to the same affect*." The Court refers to a series of declarations in the introduction to the CSA saying that (1) local distribution and possession of controlled substances causes "swelling" in interstate traffic; (2) local production and distribution cannot be distinguished from interstate production and distribution; (3) federal control over intrastate incidents "is essential to effective control" over interstate drug trafficking. These bare declarations cannot be compared to the record before the Court in *Wickard*.

They amount to nothing more than a legislative insistence that the regulation of controlled substances must be absolute. They are asserted without any supporting evidence—descriptive, statistical, or otherwise. . . .

There Is No Evidence That Medical Marijuana Laws Would Be Abused

In particular, the CSA's introductory declarations are too vague and unspecific to demonstrate that the federal statutory scheme will be undermined if Congress cannot exert power over individuals like respondents. The declarations are not even specific to marijuana. (Facts about substantial effects may be developed in litigation to compensate for the inadequacy of Congress' findings; in part because this case comes to us from the grant of a preliminary injunction, there has been no such development.) Because here California, like other States, has carved out a limited class of activity for distinct regulation, the inadequacy of the CSA's findings is especially glaring. The California Compassionate Use Act exempts from other state drug laws patients and their caregivers "who posses[s] or cultivat[e] marijuana for the *personal* medical purposes of the patient upon the written or oral recommendation of a physician" to treat a list of serious medical conditions. The Act specifies that it should not be construed to supersede legislation prohibiting persons from engaging in acts dangerous to others, or to condone the diversion of marijuana for nonmedical purposes. To promote the Act's operation and to facilitate law enforcement, California recently enacted an identification card system for qualified patients. We generally assume States enforce their laws, and have no reason to think otherwise here.

The Government has not overcome empirical doubt that the number of Californians engaged in personal cultivation, possession, and use of medical marijuana, or the amount of marijuana they produce, is enough to threaten the federal regime. Nor has it shown that Compassionate Use Act marijuana users have been or are realistically likely to be responsible for the drug's seeping into the market in a significant way. . . .

It says that the California statute might be vulnerable to exploitation by unscrupulous physicians, that Compassionate Use Act patients may overproduce, and that the history of the narcotics trade shows the difficulty of cordoning off any drug use from the rest of the market. These arguments are plausible; if borne out in fact they could justify prosecuting Compassionate Use Act patients under the federal CSA. But, without substantiation, they add little to the CSA's conclusory statements about diversion, essentiality, and market effect. Piling assertion upon assertion does not, in my view, satisfy the substantiality test of *Lopez* and *Morrison*.

We would do well to recall how James Madison, the father of the Constitution, described our system of joint sovereignty to the people of New York:

> The powers delegated by the proposed constitution to the federal government are few and defined. Those which are to remain in the State governments are numerous and indefinite. . . . The powers reserved to the several States will extend to all the objects which, in the ordinary course of af-

fairs, concern the lives, liberties, and properties of the people, and the internal order, improvement, and prosperity of the State.

Relying on Congress' abstract assertions, the Court has endorsed making it a federal crime to grow small amounts of marijuana in one's own home for one's own medicinal use. This overreaching stifles an express choice by some States, concerned for the lives and liberties of their people, to regulate medical marijuana differently. If I were a California citizen, I would not have voted for the medical marijuana ballot initiative; if I were a California legislator, I would not have supported the Compassionate Use Act. But whatever the wisdom of California's experiment with medical marijuana, the federalism principles that have driven our Commerce Clause cases require that room for experiment be protected in this case. For these reasons I dissent.

> "Despite its apparent importance to drug warriors, [Gonzales] v. Raich *is not about* . . . *the wisdom, or lack thereof, of allowing chronically ill individuals to smoke weed for medicinal purposes.*"

Gonzales v. Raich Is Less About Drugs than About Limits to Federal Power

Jonathan H. Adler

Jonathan H. Adler is an associate professor and associate director of the Center for Business Law and Regulation at the Case West-ern Reserve University School of Law.
In the following viewpoint, written before the Supreme Court made its decision in Gonzales v. Raich, *he argues that much more is at issue than drug posses-sion. The Constitution limits the power of the federal govern-ment, he says, and a ruling for the government would exceed those limits. The federal Controlled Substances Act (CSA), which prohibits the possession of even small amounts of marijuana, is based on the federal government's power to regulate interstate commerce. However, California's medical marijuana law allows it to be grown only for personal use in a completely noncommer-cial way. The government claims that it will nevertheless have an effect on commerce simply because an interstate drug market exists. Under this reasoning, Adler says, there is no limit to what the federal government might try to control. Therefore, many*

Jonathan H. Adler, "High Court High Anxiety," *National Review Online*, December 1, 2004. www.nationalreview.com. Copyright © 2004 by National Review, Inc., 215 Lexing-ton Avenue, New York, NY 10016. Reproduced by permission.

large groups have become involved in the case, and even some that are opposed to the medical use of marijuana are strongly against the overruling of California's law.

A ngel McClary Raich is seriously ill. Diagnosed with an in-operable brain tumor and several complicating condi-tions, Raich found traditional medical treatments to be of little use. Having exhausted every legal alternative, her doctor recommended that she try marijuana—and it worked. Like many individuals suffering from chronic pain or loss of appe-tite, Raich found that marijuana alleviated her symptoms sub-stantially. Yet to continue with this treatment, Raich, and those who supply her with marijuana free of charge, had to break federal law. Under California law, Raich can possess and use marijuana pursuant to a doctor's prescription or recom-mendation. Yet according to the federal government, even such minimal marijuana possession approved by a doctor re-mains illegal.

Fearing potential prosecution, Raich went to federal court seeking a declaratory judgment that, among other things, the federal government lacks the constitutional authority to pro-hibit simple marijuana possession for personal medical use. Represented by noted libertarian law professor Randy Barnett, Raich argued that, at least as applied to her situation, the fed-eral Controlled Substances Act (CSA) is unconstitutional. This week, the Supreme Court heard oral arguments in her case, *[Gonzales] v. Raich*. At stake is more than California's effort to legalize the medical use of marijuana. A decision for the fed-eral government could send federalism and the constitutional doctrine of enumerated powers up in smoke.

The Constitution Limits Federal Power

On behalf of Raich, Professor Barnett argued that the cultiva-tion and possession of marijuana "solely for the personal medical use of seriously ill individuals, as recommended by their physician and authorized by State law" is simply beyond the reach of federal power. Under our constitutional structure, states retain "broad powers to define criminal law, regulate medical practice, and protect the lives of their citizens." Federal power, on the other hand, is limited to the specific grant of enumerated powers in the Constitution, and does not reach mundane questions of criminal law. No matter how worthy the purpose of a given federal statute, it remains invalid if it exceeds the constitutionally proscribed bounds.

The federal government maintains that it has the power to prohibit the possession of any and all drugs, even in infinitesimal amounts, and therefore that California's effort to legalize medical marijuana is preempted by federal law. Under the CSA, it is a federal crime to possess "schedule I" drugs like marijuana, with or without a doctor's prescription. Like most federal regulatory statutes, the CSA was enacted pursuant to Congress's power to "regulate commerce . . . among the several states." As currently understood, this clause grants Congress the broad power to regulate commercial enterprises and other activities that have a "substantial effect" on interstate commerce. There is little question that this entails the power to regulate the production, distribution, and sale of pharmaceuticals, particularly insofar as medical markets are of national scope. Congress can empower the Food and Drug Administration [FDA] to set conditions on the sale of approved pharmaceuticals and may authorize the Drug Enforcement Administration [DEA] to arrest those who buy and sell drugs contrary to federal law.

In this case, the federal government also maintains that it can prohibit the simple possession of a drug for medical purposes, even when authorized and regulated by a validly adopted state law, and even if conducted in a wholly noncommercial fashion. Such power, the federal government asserts, is necessary to maintain a comprehensive federal regulatory system for the use and distribution of drugs. Moreover, even the mere possession of drugs can "substantially affect" interstate commerce, as there is a vibrant, albeit illegal, interstate drug market.

This argument proves too much. Under the government's reasoning there is no activity beyond Congress's grasp—a position the Supreme Court has repeatedly rejected over the past ten years. Essentially, the Justice Department maintains that the power to adopt broad economic regulatory schemes necessarily entails the power to reach the most inconsequential, noncommercial conduct that occurs wholly within the confines of a single state. Even at the height of federal power during the New Deal, the Supreme Court never authorized an assertion of federal power as expansive as is at issue here. Should the Court uphold the assertion of federal power in this case, constitutional limitations on the exertion of enumerated federal powers could well disappear.

Under the federal government's logic, Congress could enact an omnibus child-care statute, regulating the care and feeding of children and infants in private homes, because child care is often an economic enterprise and the federal government could assert an interest in regulating the market for child-care services. Not even the infamous case of *Wickard v. Filburn*, in which the Supreme Court upheld Congress's power to regulate the planting of wheat on an individual farm, reached this far. At least farmer Filburn was engaged in economic activity—planting wheat as part of a larger economic enterprise (his farm). Angel Raich's marijuana possession, however, lacks even this passing connection to economic activity. It was on this ground that the Supreme Court struck down federal statutes prohibiting gun possession in or near schools and penalizing gender-motivated violence. In neither case could the activity be remotely considered "economic" nor can the local marijuana possession at issue in *[Gonzales] v. Raich*.

The Case Is About More than Drug Possession

The importance of the case can be seen in the line-up of amicus [friend of the court] briefs supporting Raich's case. Noted conservative legal scholars, including former [President Ronald] Reagan and [President George H.W.] Bush assistant attorney general Douglas Kmiec and former Bush solicitor general Charles Fried, signed or authored briefs urging the Court to recognize that federal power cannot reach this far. Several states have done the same. Few groups not directly involved in anti-drug efforts have lined up on the Justice Department's side.

Some drug warriors fear that a victory for Raich could hamper federal anti-drug efforts because drug possession is much easier to prove than is the intent to distribute. Yet possession of small amounts of marijuana is rarely prosecuted under federal law as it is. State and local governments are responsible for most enforcement of low-level drug crimes. If petty possession needs to be prosecuted, it can be under state law. Just as the Supreme Court's invalidation of the Gun-Free School Zones Act did not produce a flood of firearms in the nation's schools, striking down the application of federal law in this case won't end marijuana prohibition. Even in California, marijuana possession for anything other than medical use remains illegal. If arrested, a medical-marijuana user bears the burden of proving that it was for medicinal use.

Alabama solicitor general Kevin Newsom filed a particularly powerful brief on behalf of several states with strong anti-drug policies, maintaining that California's medical-marijuana law poses no threat to those states, such as Alabama, where marijuana remains illegal. Alabama prosecutes drug crimes vigorously, including pot possession, and has some of the harshest drug-possession penalties in the country. Yet, Alabama and other states intervened to "support their neighbor's prerogative in our federalist system to serve as 'laboratories for experimentation.'" While agreeing with drug prohibitionists that California's medical-marijuana policy is "profoundly misguided," Alabama argued that upholding the federal prosecution of medical-marijuana users is a greater threat than letting sick people get high in California pursuant to a validly enacted state law.

From the earliest days of the Republic, the Supreme Court has emphasized that the Constitution creates a federal government of "limited and enumerated powers." There is no federal "police power" authorizing Congress to cure every injustice or right every wrong. Rather, the federal government was entrusted with those limited and discrete powers necessary for national cohesion. Matters of truly national import—matters that cannot be handled by state and local governments acting alone or in concert—are entrusted to the federal government. As made explicit in the Constitution's texts, all others powers remain in the hands of the states and the people.

Despite its apparent importance to drug warriors, *[Gonzales] v. Raich* is not about medical marijuana or drug prohibition. Nor is it about the wisdom, or lack thereof, of allowing chronically ill individuals to smoke weed for medicinal purposes. Rather, it concerns the limits of federal power under the Constitution. Federalism does not play favorites. It limits the scope of federal power to pursue liberal and conservative ends alike. If a majority of the Court remembers this lesson, Angel Raich will get to keep her medicine. More important, the nation will keep the constitutional limits on federal power.

> *"In view of the tens of thousands of potential medical marijuana users ... allowing local production and distribution for these users plainly would pose large risks to Congress' effort to ban marijuana from interstate commerce."*

Any Change in Drug Policy Should Be Made by Congress, Not the Courts

Timothy J. Dowling

Timothy J. Dowling is chief counsel of Community Rights Counsel, a Washington, D.C., public interest law firm that filed an amicus curiae (friend of the court) brief in support of the federal government in Gonzales v. Raich, *in which the Supreme Court held that federal drug laws apply even in states where medical use of marijuana is legal. In the following viewpoint, written before the Supreme Court made its ruling, Dowling argues that the question is not whether using marijuana for medical purposes should be allowed, but whether federal courts should decide such issues through a narrow interpretation of the Constitution. In Dowling's opinion, a ban on marijuana cultivation falls well within the limits of the federal government's power because control of drug traffic within a state is essential to controlling interstate drug traffic. Many federal laws such as environmental safeguards and civil rights laws are based on the authority of Congress to regulate commerce, so the legal system has a large stake in maintaining that authority. The Court cannot separate the relatively small group of people who use medical marijuana from other users, he says; it must consider the entire class of conduct to which a regulation is applicable. Furthermore, state law cannot be allowed to overrule federal law. Dowling believes that any change in laws regarding drug use should be made by Congress and not imposed by the courts.*

The central question in *[Gonzales] v. Raich*, a commerce clause challenge to federal drug laws as applied to the medical use of marijuana, is not whether to allow such use or whether scarce federal enforcement resources should be employed to seize marijuana plants from those suffering from debilitating illnesses. Rather, the question before the Supreme Court on Nov. 29 [2004] is whether federal courts should decide these policy issues through an unduly narrow reading of the commerce clause that would threaten many vital federal protections.

The vast majority of federal environmental safeguards, civil rights laws, safety standards, and other federal protections are rooted in Congress' commerce clause authority. The same is true for most federal criminal laws, including bans on heroin, methamphetamines, and other controlled substances.

As noted by Justice Anthony Kennedy in *United States v. Lopez* (1995), "the Court as an institution and the legal system as a whole have an immense stake in the stability of our commerce clause jurisprudence as it has evolved to this point." In Kennedy's words, this immense stake "counsels great restraint" in entertaining commerce clause challenges.

Comfortably Constitutional

The commerce clause, as supplemented by the necessary-and-proper clause (which grants Congress the power to make all laws "necessary and proper" to execute federal authority), has long been interpreted broadly to support the federal regulation needed for our complex society.

Time-honored precedent—reaffirmed in both *Lopez* and *United States v. Morrison* (2000)—makes clear that Congress may regulate economic conduct where it rationally concludes that the conduct, when aggregated with similar activity by others, has a significant effect on interstate commerce. Under these rulings, a court may not excise a subclass of conduct from otherwise legitimate regulation simply because the impact of the subclass is small.

The federal ban on local marijuana production and distribution at issue in *Raich* falls comfortably within these constitutional bounds. Congress rationally found that local production and distribution of marijuana, heroin, and other controlled substances swell interstate drug traffic. And Congress reasonably concluded that it is impractical to differentiate the controls applying to *intrastate* and *interstate* drug traffic, and that regulation of intrastate drug traffic is "essential" to the effective control of the interstate drug market.

These determinations might be debatable, but they cannot be dismissed as irrational, and they should make *Raich* an easy case.

The U.S. Court of Appeals for the 9th Circuit sidestepped Congress' findings, however, by narrowly redefining the relevant class of activity as "the intrastate, noncommercial cultivation, possession and use of marijuana for personal medical purposes on the advice of a physician." It then concluded that this redefined class is not economic and thus not subject to aggregation, and that the specific marijuana use at issue in *Raich* has no substantial effect on interstate commerce.

The lower court's errors are twofold. First, as is evident from any economics text, local cultivation and distribution of marijuana, even without a sale, is economic activity. In both *Lopez* and *Morrison*, the Court described the regulated conduct in *Wickard v. Filburn* (1942) as the "production and consumption of home-grown wheat." It concluded that this activity is economic and properly aggregated with similar conduct by others in analyzing the impact on interstate commerce.

Remember Ollie's Barbecue

Second, and more fundamentally, the lower court's redefinition of the regulated class ignores the bedrock principle that Congress may define and regulate a class of conduct where the definition is rational and the class substantially affects interstate commerce.

For instance, when the Supreme Court upheld the federal ban on racial discrimination by restaurants in *Katzenbach v. McClung* (1964), it did not allow the restaurant challenging the law, Ollie's Barbecue, to redefine the relevant class to include only small, family-owned restaurants in Birmingham, Ala., that had no record of serving interstate travelers.

Instead, the Court considered whether discrimination at the much broader class of restaurants covered by our civil rights laws could be viewed as substantially affecting interstate commerce. The lower court's analysis in *Raich* contravenes *Katzenbach* and decades of similar precedents.

Likewise, federal regulation of loan sharking, wetlands destruction, and other site-specific activity is not invalid simply because it is possible to redefine the regulated class more narrowly so that the impact appears small. These and many more federal protections would be at risk under the lower court's analysis.

Nor does it matter that California law permits medical marijuana use. The scope of congressional power does not turn on the vagaries of state law. Indeed, allowing state law to trump federal authority would turn the supremacy clause on its head.

To be sure, there are limits on Congress' authority over non-economic activity, as shown by *Lopez* and *Morrison*, which invalidated federal bans on mere gun possession near schools and gender-related violence. But these cases are easily distinguished from regulation of economic activity, such as drug production and distribution, which must be considered in the aggregate. In view of the tens of thousands of potential medical marijuana users in California and elsewhere, allowing local production and distribution for these users plainly would pose large risks to Congress' effort to ban marijuana from interstate commerce.

Any change to the federal drug statutes should be made by Congress. It should not be forced on the nation by the courts through a cramped reading of the Constitution that threatens federal protections across the board.

*"A telephone survey of a random na-
tional sample of registered voters . . .
indicated that 68 percent of respondents
opposed federal prosecution of patients
who use marijuana for medical reasons."*

The Battle Between
Federal and State
Governments over Medical
Marijuana Is Not Over

Susan Okie

Susan Okie is a contributing editor of the New England Journal
of Medicine. *In the following viewpoint, she discusses the status
of medical marijuana after the Supreme Court ruled that its
possession is illegal under federal law even where it is legal under state law. States that allow it have not changed their policies,
although some have cracked down on abuse. It is not clear to
what extent the federal government will arrest and prosecute
medical users, says Okie, but patients with diseases that are
helped by marijuana are still using it on the advice of their doctors. Research is being conducted into marijuana's medical benefits, although this is slow because it is difficult for researchers to
obtain a legal supply. The federal government still maintains
that marijuana has no medical value and Congress has refused
to pass legislation prohibiting the use of federal funds to prosecute people who use it for medicinal purposes. However, advocates are working to get medical use approved by more states,*

and in Okie's opinion, the battle between federal and state governments is far from over.

Angel McClary Raich, a California woman at the center of the recent Supreme Court case on medical marijuana, hasn't changed her treatment regimen since the Court ruled in June [2005] that patients who take the drug in states where its medicinal use is legal are not shielded from federal prosecution. A thin woman with long, dark hair and an intense gaze, Raich takes marijuana, or cannabis as she prefers to call it, about every two waking hours—by smoking it, by inhaling it as a vapor, by eating it in foods, or by applying it topically as a balm. She says that it relieves her chronic pain and boosts her appetite, preventing her from becoming emaciated because of a mysterious wasting syndrome. Raich and her doctor maintain that without access to the eight or nine pounds of privately grown cannabis that she consumes each year, she would die.

Although Raich has embraced a public role advocating the medicinal use of marijuana, she says that her health suffered during the hectic days following the announcement of the Court's decision, when a whirlwind schedule of press conferences and congressional meetings in Washington prevented her from medicating herself with cannabis as regularly as she needed to. "My body was shutting down on me," she said in an interview from her Oakland home last month. "I'm scared of my health failing. I'm scared of the federal government coming in and doing more harm. [Recently,] the city of Oakland warned there were going to be some raids" on marijuana dispensaries. "We're all just waiting. Sitting on the frontline is extremely stressful."

In the Supreme Court case *Gonzales v. Raich*, the justices ruled 6 to 3 that the federal government has the power to arrest and prosecute patients and their suppliers even if the marijuana use is permitted under state law, because of its authority under the federal Controlled Substances Act to regulate interstate commerce in illegal drugs. In practical terms, it is not yet clear what effect the Court's decision will have on patients. An estimated 115,000 people have obtained recommendations for marijuana from doctors in the 10 states that have legalized the cultivation, possession, and use of marijuana for medicinal purposes. Besides California, those states are Alaska, Colorado, Hawaii, Maine, Montana, Nevada, Oregon, Vermont, and Washington. (Three weeks after the decision was announced, Rhode Island's legislature passed a similar law and soon afterward overrode a veto by the state's governor.)

Use of Medical Marijuana Is Still Legal Under State Laws

Immediately after the news of the high court's ruling, attorneys general in the states that have approved the use of medical marijuana emphasized that the practice remained legal under their state laws, and a telephone survey of a random national sample of registered voters, commissioned by the Washington-based Marijuana Policy Project, indicated that 68 percent of respondents opposed federal prosecution of patients who use marijuana for medical reasons. Nationally, most marijuana arrests are made by state and local law-enforcement agencies, with federal arrests accounting for only about 1 percent of cases. However, soon after the decision was announced, federal agents raided three of San Francisco's more than 40 medical marijuana dispensaries. Nineteen people were charged with running an international drug ring; they allegedly were using the dispensaries as a front for trafficking in marijuana and in the illegal amphetamine "ecstasy."

In California, the raids were widely viewed as a signal that federal drug enforcement agents intended to crack down on abuse of the state's medical marijuana program. California has an estimated 100,000 medical marijuana users. Its 1996 law grants doctors much greater latitude in recommending the drug than do similar laws in other states, and the U.S. District Court for the Northern District of California ruled in 2000 that doctors who prescribe marijuana are protected from federal prosecution under the First Amendment, provided that they do not help their patients obtain the drug. In San Francisco, some journalists or investigators who posed as patients have reported that they had little difficulty obtaining a recommendation for medical marijuana, which allows the holder to purchase the drug from a dispensary. "We're empathetic to the sick," the Drug Enforcement Administration's Javier Pena told reporters after the raids, "but we can't disregard the federal law."

Even before the Supreme Court decision, many Californians had been calling for stricter state regulation of medical marijuana. Some cities have banned marijuana dispensaries, and many counties and cities—including San Francisco—have imposed moratoriums on the opening of new ones. Some local jurisdictions register and issue identification cards to patients who use marijuana for medical reasons, and state officials have been working on a voluntary statewide registration program. However, the officials recently put the program on hold, citing concern that the issuance of identification cards to patients might put state health officials at risk of prosecution for aiding a federal crime and that federal drug-enforcement agents might seek state records in order to identify medical marijuana users. Registration of patients and the issuance of identification cards by the state are required in seven other states that have legalized the medical use of marijuana; patients can show the card as a defense against arrest by local or state police for possession of the drug. Maine and Washington do not issue identification cards to patients.

Conditions for which marijuana is commonly recommended include nausea caused by cancer chemotherapy; anorexia or wasting due to cancer, AIDS, or other diseases; chronic pain; spasticity caused by multiple sclerosis or other neurologic disorders; and glaucoma. Frank Lucido, a Berkeley family practitioner who is Raich's doctor, said that so far, the Court ruling appears to have had little effect on his patients who use medical marijuana. About 30 percent of Lucido's practice consists of evaluating patients who want a recommendation for the drug. He said in an interview that he will not issue such a recommendation unless a patient has a primary care physician and has a condition serious enough to require follow-up at least annually. About 80 percent of his patients who use medical cannabis have chronic pain; a smaller number take the drug for muscle spasms, mood disorders, migraine, AIDS, or cancer. "My patients probably average in their 30s," Lucido said. "I have had probably five patients who are under 18. These are people with serious illnesses, where parents were very clear that this would be a good medication for them."

Peter A. Rasmussen, an oncologist in Salem, Oregon, said he discusses the option of trying marijuana with about one in ten patients in his practice. "It's not my first choice for any symptom," he said in an interview. "I only talk about it with people if my first-line treatment doesn't work." Rasmussen said marijuana has helped stimulate appetite or reduce nausea in a number of his patients with cancer, but others have been distressed by its psychological effects. Some express interest in trying marijuana but have difficulty getting the drug. "Most of my patients who use it, I think, just buy the drug illegally," he said. "But a lot of my patients, they're older, they don't know any kids, they don't hang out on the street. They just don't know how to get it."

Research on Medical Marijuana Is Continuing

Clinical research on marijuana has been hampered by the fact that the plant, which contains dozens of active substances, is an illegal drug classified as having no legitimate medical use. Researchers wishing to do clinical studies must first get government permission and obtain a supply of the drug from the National Institute on Drug Abuse. In a report published in 1999, an expert committee of the Institute of Medicine expressed concern about the adverse health effects of smoking marijuana, particularly on the respiratory tract. The report called for expanded research on marijuana's active components, known as cannabinoids, including studies to explore the chemicals' potential therapeutic effects and to develop safe, reliable, rapid-onset delivery systems. It also recommended short-term clinical trials of marijuana "in patients with conditions for which there is reasonable expectation of efficacy."

There has been some progress toward those goals. The Center for Medicinal Cannabis Research (CMCR), a three-year research initiative established in 1999 by the California state legislature, has funded several placebo-controlled clinical trials of smoked marijuana to treat neuropathic pain, pain from other causes, and spasticity in multiple sclerosis, and the results are likely to be available soon. The National Institute on Drug Abuse provided both the active marijuana and the "placebo," a smokable version of the drug from which dronabinol (Δ9-tetrahydrocannabinol, or THC) and certain other active constituents had been removed. "It's like decaf coffee or nicotine-free cigarettes, and it tastes the same [as marijuana]," said Igor Grant, a professor of psychiatry at the University of California, San Diego, and director of the CMCR. He said additional studies of the whole plant, as well as its individual components, are still needed. "It's still the case that we don't know which components of botanical marijuana have beneficial effects, if any," he said.

In an open-label trial, oncologist Donald I. Abrams of the University of California, San Francisco, found evidence of marijuana's effectiveness in the treatment of neuropathic pain among HIV-infected patients and has just finished a placebo-controlled trial that he intends to publish soon. Abrams has also shown that cannabinoids that are smoked or taken orally do not adversely affect drug treatment of HIV, and he is completing a study that compares blood levels of cannabinoids among volunteers who inhaled vaporized marijuana with similar levels among volunteers who smoked the drug. Vaporizers heat the drug to a temperature below that required for combustion, producing vapor that contains the active ingredients without the tar or particulates thought to be responsible for most of the drug's adverse effects on the respiratory tract.

Meanwhile, a new marijuana-derived drug is on the Canadian market and may soon be considered for approval by the Food and Drug Administration. Sativex, a liquid cannabis extract that is sprayed under the tongue, was approved in Canada in June [2005] for the treatment of neuropathic pain in multiple sclerosis. Its principal active ingredients are dronabinol and cannabidiol, which are believed to be the primary active components of marijuana. The drug's manufacturer, GW Pharmaceuticals of Britain, is also testing it for cancer pain, rheumatoid arthritis, postoperative pain, and other indications. Marinol, a synthetic version of dronabinol supplied in capsules, is approved in the United States for chemotherapy-associated nausea and for anorexia and wasting among patients with AIDS.

The Battle Between Federal and State Governments Is Not Over

On the day the Supreme Court ruling was announced [June 6, 2005], John Walters, President George W. Bush's "drug czar," issued a statement declaring, "Today's decision marks the end of medical marijuana as a political issue. . . . We have a responsibility as a civilized society to ensure that the medicine Americans receive from their doctors is effective, safe, and free from the pro-drug politics that are being promoted in America under the guise of medicine." Nine days later, the House of Representatives, for the third year in a row, defeated a measure that would have prevented the Justice Department from spending money to prosecute medical marijuana cases under federal law.

Nevertheless, marijuana advocates insist that the long-running battle between federal and state governments over the medicinal use of marijuana is far from over. Activists next plan to focus on getting more states to pass laws legalizing medical marijuana, according to Steve Fox, former director of government relations for the Marijuana Policy Project.

It is surprising that the Supreme Court decision does not necessarily spell the end even of Angel Raich's legal case. Raich and another California patient, Diane Monson, who initially sued to prevent the Justice Department from prosecuting them or their suppliers, won a favorable ruling in 2003 from California's Court of Appeals for the Ninth Circuit. The Supreme Court's reversal now sends their case back to that court. Raich said that she, Monson, and their attorneys will ask the appeals court judges to consider other legal arguments, such as whether prosecuting patients who use marijuana to relieve pain violates their right to due process of law. "Previous decisions have established that there is a fundamental right to preserve one's life and avoid needless pain and suffering," explained Boston University's Randy Barnett, a constitutional lawyer who argued the women's case before the Supreme Court. "Federal restriction on accessibility to medical cannabis is an infringement" on that right, he said.

Raich vowed to continue her personal battle. "I'm stubborn as heck, so I don't plan to give it up that easily. I plan to fight until I can't fight anymore," she said.

Organizations to Contact

The editors have compiled the following list of organizations concerned with the issues debated in this book. The descriptions are derived from materials provided by the organizations. All have publications or information available for interested readers. The list was compiled on the date of publication of the present volume; the information provided here may change. Be aware that many organizations take several weeks or longer to respond to inquiries, so allow as much time as possible.

Angel Wings Patient Outreach
4100 Redwood Road, #365, Oakland, CA   94619-2363
E-mail: angelwings@angeljustice.org
Web site: www.angeljustice.org

Angel Wings is a nonprofit corporation that focuses its resources throughout California on activities most likely to have an impact on clarifying and vindicating the civil rights, human rights, and disability rights of medical cannabis patients, their caregivers, and their physicians. Its Web site contains detailed information about Angel Raich and the court case *Gonzales v. Raich.*

Common Sense for Drug Policy (CSDP)
377-C Spencer Avenue, Lancaster, PA   17603
(717) 299-0600 • Fax: (717) 393-4953
E-mail: info@csdp.org
Web site: www.csdp.org

Common Sense for Drug Policy (CSDP) is a nonprofit organization dedicated to reforming drug policy and expanding harm reduction. It disseminates factual information and comments on existing laws, policies and practices. Its Web site contains news and an extensive list of links to other organizations, both those that oppose current policy and those that favor it. In addition to its main Web site, it maintains Drug War Facts (www.drugwarfacts.org), Drug War Distortions (www.drugwardistortions.org), and the Coalition for Medical Marijuana (www.medicalmj.org), all of which offer extensive information.

Criminal Justice Policy Foundation (CJPF)
8730 Georgia Avenue, Suite 400, Silver Spring, MD   20910
(301) 589-6020 • Fax: (301) 589-5056
E-mail:info@cjpf.org
Web site: www.cjpf.org

The Criminal Justice Policy Foundation (CJPF) is a private nonprofit organization that works to educate the public about the impact of drug policies and the problems of policing the criminal justice system. Its Web site contains many articles and links to resources dealing with the war on drugs.

Drug Action Network
E-mail: info@DrugActionNetwork.com
Web site: www.DrugActionNetwork.com

The Drug Action Network is a nonprofit group composed of parents, teachers, students, and concerned citizens who believe that the war on drugs has made the problems of drug use far worse than they were or should be by creating a black market. Its Web site contains a detailed history of the drug war and information about drug laws, as well as articles arguing that prohibition has done harm.

Drug Enforcement Administration (DEA)
Mailstop AES, 8701 Morrissette Drive, Springfield, VA   22152
(202) 307-1000
Web site: www.usdoj.gov/dea

The Drug Enforcement Administration (DEA) is the agency responsible for enforcing the controlled substances laws and regulations of the United States. Its Web site contains information about its activities as well as speeches and testimony in favor of the war on drugs.

Drug Free America Foundation (DFAF)
2600 Ninth Street N, Suite 200
St. Petersburg, FL   33704-2744
(727) 828-0211 • Fax: (727) 828-0212
E-mail: librarian@dfaf.org
Web site: www.dfaf.org

Drug Free America Foundation (DFAF) is a nongovernmental drug prevention and policy organization committed to developing, promoting and sustaining global strategies, policies and laws that will reduce illegal drug use. It favors the war on drugs, and its Web site contains many articles defending current policy, including student drug testing.

Drug Policy Alliance (DPA) Network
70 West Thirty-Sixth Street, Sixteenth Floor
New York, NY   10018
(212) 613-8020 • Fax: (212) 613-8021
Web site: www.drugpolicy.org

The Drug Policy Alliance (DPA) Network is the nation's leading organization promoting policy alternatives to the drug war that are grounded in science, compassion, health and human rights. Its mission is to advance policies and attitudes that best reduce the harms of both drug misuse and drug prohibition, and to promote the sovereignty of individuals over their minds and bodies. Its Web site offers a large searchable library of on-line documents related to the war on drugs as well as a catalog of over 15,000 books held in its New York library.

Drug Reform Coordination Network (DRCNet)
1623 Connecticut Avenue NW, Third Floor
Washington, DC   20009
(202) 293-8340 • Fax: (202) 293-8344
E-mail: drcnet@drcnet
Web site: www.drcnet.org

The Drug Reform Coordination Network (DRCNet) is an activist organization that advocates an end to drug prohibition and its replacement with some sensible framework in which drugs can be regulated and controlled. It publishes the widely used weekly newsletter *Drug War Chronicle*, which is available at its Web site with a topical index. It also maintains the world's largest online library of drug policy (www.druglibrary. org).

DrugSense/MAP
14252 Culver Drive #328, Irvine, CA   92604-0326
(800) 266-5759
E-mail: info@drugsense.org
Web site: www.drugsense.org

DrugSense and its largest project MAP constitute a nonprofit corporation that exists to provide accurate information relevant to drug policy in order to "heighten awareness of the extreme damage being caused to our nation and the world by our current flawed and failed war on drugs." It aims to inform the public of the existence of rational alternatives to the drug war, and to help organize citizens to bring about needed reforms. Its Web site contains many articles and links to resources offered by other drug reform organizations.

Educators for Sensible Drug Policy (EFSDP)
E-mail: info@efsdp.org
Web site: www.efsdp.org

Educators for Sensible Drug Policy's aim is to encourage drug use to be treated as a health issue and not a criminal behavior. Its Web site contains resources on many drug issues, including school drug testing and medical marijuana use, among others.

Forfeiture Endangers American Rights (FEAR)
20 Sunnyside, Suite A-419, Mill Valley, CA   94941
(888) 332-7001
Web Site: www.fear.org

Forfeiture Endangers American Rights (FEAR) is a nonprofit organization focused on unfair forfeiture laws and the harm they do to innocent people. In addition to legal information and the stories of victims in non-drug related cases, its Web site includes many articles involving the war on drugs, plus audio of the Supreme Court's announcement of its decision in *United States v. Ursery* (1996).

Institute for a Drug-Free Workplace
8614 Westwood Center Drive, Suite 950, Vienna, VA   22182
(703) 288-4300
Web site: www.drugfreeworkplace.org

The Institute for a Drug-Free Workplace is dedicated to educating employers, employees, state and federal legislators, and the public at large about the impact of drug abuse on the workplace. Such topics discussed include the rights and responsibilities of employers and employees; the status of legislative, regulatory, and legal developments at the state and federal levels; and the role of employers in the national effort to combat drug abuse. The Institute publishes extensive materials on these subjects, much of which are available for purchase. Its Web site offers detailed results of a Gallup poll on what employees think about drug abuse and drug testing.

Law Enforcement Against Prohibition (LEAP)
121 Mystic Avenue, Medford, MA   02155
(781) 393-6985 • Fax: (781) 393-2964
E-mail: info@leap.cc
Web site: www.leap.cc

Law Enforcement Against Prohibition (LEAP) is a nonprofit educational organization founded by police officers. Its mission is to reduce the multitude of unintended harmful consequences resulting from fighting the war on drugs. Its Web site contains many articles and multimedia presentations explaining why it believes legalizing drugs would be a more effective way of reducing the crime, disease, and addiction drugs cause.

National Organization for the Reform
of Marijuana Laws (NORML)

1600 K Street NW, Suite 501, Washington, DC   20006-2832
(888) 676-6765 • Fax: (202) 483-0057
E-mail: norml@norml.org
Web site: www.norml.org

The National Organization for the Reform of Marijuana Laws' mission is "to move public opinion sufficiently to achieve the repeal of marijuana prohibition so that the responsible use of cannabis by adults is no longer subject to penalty." Its Web site contains considerable information on medical use of marijuana, pre-employment drug testing, and student drug testing, including the Supreme Court cases involving these issues.

Office of National Drug Control Policy (ONDCP)

PO Box 6000, Rockville, MD   20849-6000
(800) 666-3332 • Fax: (301) 519-5212
Web site: www.whitehousedrugpolicy.gov

The White House Office of National Drug Control Policy is a component of the Executive Office of the President; it is headed by the federal "drug czar." Its principal purpose is to establish policies, priorities, and objectives for the nation's drug control program. Its Web site offers the National Drug Control Strategy document, press releases, fact sheets, and links, plus many publications about drug war policy that can be downloaded.

For Further Research

Books

Nelson V. Arcos, *The War on Drugs: Winning, Losing or Treading Water*. Hauppauge, NY: Nova Science, 2005.

David Sadofsky Baggins, *Drug Hate and the Corruption of American Justice*. Westport, CT: Praeger, 1998.

Dan Baum, *Smoke and Mirrors: The War on Drugs and the Politics of Failure*. Boston, MA: Little, Brown, 1996.

Arthur Benavie, *Drugs: America's Holy War*. New York: Routledge, 2008.

Eva Bertram, *Drug War Politics: The Price of Denial*. Berkeley: University of California Press, 1996.

Pablo De Greiff, *Drugs and the Limits of Liberalism: Moral and Legal Issues*. Ithaca, NY: Cornell University Press, 1999.

Charles Doyle, *Crime and Forfeiture*. Hauppauge, NY: Nova Science, 2008.

Dirk Chase Eldridge, *Ending the War on Drugs: A Solution for America*. Bridgehampton, NY: Bridge Works, 1998.

Gary L. Fisher, *Rethinking Our War on Drugs: Candid Talk About Controversial Issues*. Westport, CT: Praeger, 2006.

James P. Gray, *Why Our Drug Laws Have Failed and What We Can Do About It: A Judicial Indictment of the War on Drugs*. Philadelphia: Temple University Press, 2001.

Mike Gray, *Drug Crazy: How We Got into This Mess and How We Can Get Out*. New York: Routledge, 2000.

Richard Hammersley, *Drugs and Crime: Theories and Practices*. Malden, MA: Polity Press, 2008.

Laura E. Huggins, *Drug War Deadlock: The Policy Battle Continues*. Stanford, CA: Hoover Institution Press, 2005.

Leonard W. Levy, *A License to Steal: The Forfeiture of Property*. Berkeley: University of California Press, 1995.

Bill Masters, *Drug War Addiction*. St. Louis, MO: Accurate Press, 2001.

———, ed., *The New Prohibition: Voices of Dissent Challenge the Drug War*. St. Louis, MO: Accurate Press, 2004.

Lorraine Green Mazerolle and Janice A. Roehl, eds., *Civil Remedies and Crime Prevention*. Monsey, NY: Criminal Justice Press, 1998.

Robert J. McCoun, *Drug War Heresies: Learning from Other Vices, Times, and Places*. Cambridge, MA: Cambridge University Press, 2000.

Joel Miller, *Bad Trip: How the War Against Drugs Is Destroying America*. Nashville, TN: WND Books, 2004.

Richard Lawrence Miller, *Drug Warriors and Their Prey: From Police Power to Police State*. Westport, CT: Praeger, 1996.

Jeffrey A. Miron, *Drug War Crimes: The Consequences of Prohibition*. Oakland, CA: Independent Institute, 2004.

James L. Nolan, *Reinventing Justice: The American Drug Court Movement*. Princeton, NJ: Princeton University Press, 2003.

Steve Otto, *War on Drugs or War on People? A Resource Book for the Debate*. Las Colinas, TX: Ide House, 1996.

Doris Marie Provine, *Unequal Under Law: Race in the War on Drugs*. Chicago, IL: University of Chicago Press, 2007.

Matthew B. Robinson and Renee G. Scherlen, *Lies, Damned Lies, and Drug War Statistics: A Critical Analysis of Claims Made by the Office of National Drug Control Policy*. Albany: State University of New York Press, 2007.

Thomas C. Rowe, *Federal Narcotics Laws and the War on Drugs: Money Down a Rat Hole*. New York: Routledge, 2006.

Douglas Valentine, *The Strength of the Wolf: The Secret History of America's War on Drugs*. New York: Verso, 2004.

William Weir, *In the Shadow of the Dope Fiend: America's War on Drugs*. North Haven, CT: Archon, 1997.

Steven Wisotsky, *Beyond the War on Drugs: Overcoming a Failed Public Policy*. Amherst, NY: Prometheus Books, 1990.

Periodicals

Radley Balko, "30 Years of Failure," *Reason*, June 2008.

Doug Bandow, "Medical Marijuana Not an Enemy in the Outrageous War on Drugs," *Investor's Business Daily*, December 7, 2004.

Mark Boal, "The Supreme Court vs. Teens," *Rolling Stone*, June 6, 2002.

Lawrence D. Bobo and Victor Thompson, "Unfair by Design: The War on Drugs, Race, and the Legitimacy of the Criminal Justice System," *Social Research*, Summer 2006.

Steve Brooks, "Where We Are in the Drug War," *American Legion*, March 2007.

Mary Carmichael, "In Court, Special After-School Activities," *Newsweek*, April 1, 2002.

Stefan D. Cassella, "The Case for Civil Forfeiture," 25th International Symposium on Economic Crime, September 7, 2007. http://works.bepress.com.

Peter Cassidy, "Pee First, Ask Questions Later," *In These Times*, December 20, 2002. www.inthesetimes.com.

Fern Schumer Chapman, "The Ruckus over Medical Testing," *Fortune*, August 19, 1985.

David T. Cook, Marshall Ingwerson, et al., "Stop Civil Forfeiture Abuses," *Christian Science Monitor*, June 29, 1999.

Alex Coolman, "*Gonzales v. Raich* One Year Later: Federalism Only When It's Convenient," June 5, 2006. http://druglaw.typepad.com.

Ellis Cose, "The Harm of 'Get Tough' Policies," *Newsweek*, December 13, 2007. www.newsweek.com.

Judith Wagner Decew, "Drug Testing: Balancing Privacy and Public Safety," *Hastings Center Report*, vol. 24, 1994.

Lawrence O. Gostin, "Medical Marijuana, American Federalism, and the Supreme Court," *Journal of the American Medical Association*, August 17, 2005.

Linda Greenhouse, "Court Backs Tests of Some Workers to Deter Drug Use," *New York Times*, March 22, 1989.

———, "Justices Uphold Civil Forfeiture as Anti-Drug Tool," *New York Times*, June 25, 1996.

———, "Supreme Court Seems Ready to Extend School Drug Tests," *New York Times*, March 20, 2002.

Ryan Grim, "A Guide to *Gonzales vs. Raich*," June 7, 2005. www.salon.com.

Phil Harvey and Mike Tidwell, "Let's Just Say 'No' to U.S. Drug-War Justice," *Christian Science Monitor*, August 3, 1998.

Dana Hawkins, "Trial by Vial," *U.S. News & World Report*, May 31, 1999.

Reynolds Holding, "Whatever Happened to Drug Testing?" *Time*, July 7, 2006.

Adam Liptak, "Medical Marijuana Wins a Court Victory," *New York Times*, October 30, 2002.

Dahlia Lithwick, "Urine Trouble—Uncle Sam Wants *You* to Pee in a Cup," *Slate*, March 19, 2002. www.slate.com.

S.M. Oliva, "When Did 'School Testing' Mean Testing Kids for Drugs?" June 27, 2002. www.capitalismcenter.org.

Kelly Patricia O'Meara, "When Feds Say Seize and Desist," *Insight on the News*, August 7, 2000.

Joe Pastoor, "The Mortarboard, the Sheepskin and the Dixie Cup," *Rake*, June 2004. www.rakemag.com.

Robert Pear, "Drug Testing for Customs Agents Is Abandoned by the White House," *New York Times*, January 17, 1987.

Ralph R. Reiland, "The Dirty Side of the Drug War," *Humanist*, September/October 2005.

Michael Rogers, Patricia Sellers, et al., "War on Drugs," *Fortune*, September 29, 1986. http://money.cnn.com.

Alain L. Sanders, "A Boost for Drug Testing," *Time*, April 3, 1989.

Jeff A. Schnepper, "Mandated Morality Leads to Legalized Theft," *USA Today*, March 1994.

Deborah Small, "The War on Drugs Is a War on Racial Justice," *Social Research*, Fall 2001.

Andy Sullivan, "More Drug Tests in Schools Urged by White House," *Boston Globe*, March 20, 2006.

Jacob Sullum, "Urine—or You're Out," *Reason*, November 2002.

Patrick Thomas, "Prosecuting the U.S. Drug War: Hard Policy Versus Soft Politics," *Los Angeles Times*, August 20, 1989.

Peter Vilbig, "Expanded Drug Tests in Schools? The Supreme Court Considers Whether More Students Should Be Checked," *New York Times Upfront*, May 6, 2002.

Ben Wallace-Wells and Eric Magnuson, "How America Lost the War on Drugs," *Rolling Stone*, December 13, 2007.

Mark Walsh and Linda Jacobson, "Supreme Court Allows Expansion of Schools' Drug-Testing Policies," *Education Week*, July 10, 2002.

Henry Weinstein, "To Some Jurists, High Court Ruling Brings Vindication," *Los Angeles Times*, December 11, 2007.

Index